Informing the legislative debate since 1914

Navy Ohio Replacement (SSBN[X]) Ballistic Missile Submarine Program: Background and Issues for Congress

Ronald O'Rourke
Specialist in Naval Affairs

April 8, 2014

Congressional Research Service

7-5700

www.crs.gov

R41129

Summary

The Navy's proposed FY2015 budget requests $1,219.3 million for continued research and development work on the Ohio replacement program (ORP), a program to design and build a new class of 12 ballistic missile submarines (SSBNs) to replace the Navy's current force of 14 Ohio-class SSBNs. The Ohio replacement program is also known as the SSBN(X) program. The Navy has identified the Ohio replacement program as its top program priority.

Under the Navy's FY2012 budget, the first Ohio replacement boat was scheduled to be procured in FY2019, and Ohio replacement boats were to enter service on a schedule that would maintain the Navy's SSBN force at 12 boats. The Navy's FY2013 budget deferred the procurement of the first Ohio replacement boat by two years, to FY2021. As a result of the deferment of the procurement of the lead boat from FY2019 to FY2021, the Navy's SSBN force will drop to 11 or 10 boats for the period FY2029-FY2041. The Navy says the decline to 11 or 10 boats during this period will be acceptable in terms of meeting strategic nuclear deterrent mission requirements because none of the 11 or 10 boats during that period will be encumbered by lengthy maintenance actions.

The Navy in May 2013 estimated the procurement cost of the lead ship in the program at $12.0 billion in constant 2013 dollars, including $4.6 billion in detailed design and nonrecurring engineering (DD/NRE) costs for the entire class, and $7.4 billion in construction costs for the ship itself. The Navy in April 2014 estimated the average procurement cost of boats 2 through 12 in the Ohio replacement program at about $5.36 billion each in FY2010 dollars, and is working to reduce that figure to a target of $4.9 billion each in FY2010 dollars. Even with this cost-reduction effort, observers are concerned about the impact the Ohio replacement program will have on the Navy's ability to procure other types of ships at desired rates in the 2020s and early 2030s.

Potential oversight issues for Congress for the Ohio replacement program include the following:

- the possibility that the program might experience a six-month delay due to a shortfall in FY2014 funding for manufacturing the lead ship's reactor core;

- the likelihood that the Navy will be able to reduce the average procurement cost of boats 2-12 in the program to the target figure of $4.9 billion each in FY2010 dollars;

- the accuracy of the Navy's estimate of the procurement cost of each SSBN(X);

- the prospective affordability of the Ohio replacement program and its potential impact on funding available for other Navy shipbuilding programs; and

- the question of which shipyard or shipyards will build SSBN(X)s.

This report focuses on the Ohio replacement program as a Navy shipbuilding program. CRS Report RL33640, *U.S. Strategic Nuclear Forces: Background, Developments, and Issues*, by Amy F. Woolf, discusses the SSBN(X) as an element of future U.S. strategic nuclear forces in the context of strategic nuclear arms control agreements.

Contents

Figures

Tables

Appendixes

Contacts

Introduction

This report provides background information and potential oversight issues for Congress on the Ohio replacement program (ORP), a program to design and build a new class of 12 ballistic missile submarines (SSBNs) to replace the Navy's current force of 14 Ohio-class SSBNs. The Ohio replacement program is also known as the SSBN(X) program. The Navy has identified the Ohio replacement program as its top program priority.

The Navy's proposed FY2015 budget requested $1,219.3 million for continued research and development work on the Ohio replacement program. Decisions that Congress makes on the Ohio replacement program could substantially affect U.S. military capabilities and funding requirements, and the U.S. shipbuilding industrial base.

This report focuses on the Ohio replacement program as a Navy shipbuilding program. Another CRS report discusses the SSBN(X) as an element of future U.S. strategic nuclear forces in the context of strategic nuclear arms control agreements.[1]

Background

U.S. Navy SSBNs in General

Mission of SSBNs

The U.S. Navy operates three kinds of submarines—nuclear-powered attack submarines (SSNs), nuclear-powered cruise missile submarines (SSGNs), and nuclear-powered ballistic missile submarines (SSBNs).[2] The SSNs and SSGNs are multi-mission ships that perform a variety of peacetime and wartime missions.[3] They do not carry nuclear weapons.[4]

---

[1] CRS Report RL33640, *U.S. Strategic Nuclear Forces: Background, Developments, and Issues*, by Amy F. Woolf.

[2] In the designations SSN, SSGN, SSBN, and SSBN(X), the SS stands for submarine, N stands for nuclear-powered (meaning the ship is powered by a nuclear reactor), G stands for guided missile (such as a cruise missile), B stands for ballistic missile, and (X) means the design of the ship has not yet been determined.

As shown by the "Ns" in SSN, SSGN, and SSBN, all U.S. Navy submarines are nuclear-powered. Other navies operate non-nuclear powered submarines, which are powered by energy sources such as diesel engines. A submarine's use of nuclear or non-nuclear power as its energy source is not an indication of whether it is armed with nuclear weapons—a nuclear-powered submarine can lack nuclear weapons, and a non-nuclear-powered submarine can be armed with nuclear weapons.

[3] These missions include covert intelligence, surveillance, and reconnaissance (ISR), much of it done for national-level (as opposed to purely Navy) purposes; covert insertion and recovery of special operations forces (SOF); covert strikes against land targets with the Tomahawk cruise missiles; covert offensive and defensive mine warfare; anti-submarine warfare (ASW); and anti-surface ship warfare. The Navy's four SSGNs, which are converted former SSBNs, can carry larger numbers of Tomahawks and SOF personnel than can the SSNs. SSGN operations consequently may focus more strongly on Tomahawk and SOF missions than do SSN operations. For more on the Navy's SSNs and SSGNs, see CRS Report RL32418, *Navy Virginia (SSN-774) Class Attack Submarine Procurement: Background and Issues for Congress*, by Ronald O'Rourke, and CRS Report RS21007, *Navy Trident Submarine Conversion (SSGN) Program: Background and Issues for Congress*, by Ronald O'Rourke.

[4] The Navy's non-strategic nuclear weapons—meaning all of the service's nuclear weapons other than submarine-launched ballistic missiles (SLBMs)—were removed from Navy surface ships and submarines under a unilateral U.S. (continued...)

The SSBNs, in contrast, perform a specialized mission of strategic nuclear deterrence. To perform this mission, SSBNs are armed with submarine-launched ballistic missiles (SLBMs), which are large, long-range missiles armed with multiple nuclear warheads. SSBNs launch their SLBMs from large-diameter vertical launch tubes located in the middle section of the boat.[5] The SSBNs' basic mission is to remain hidden at sea with their SLBMs, so as to deter a nuclear attack on the United States by another country by demonstrating to other countries that the United States has an assured second-strike capability, meaning a survivable system for carrying out a retaliatory nuclear attack.

Navy SSBNs, which are sometimes referred to informally as "boomers,"[6] form one leg of the U.S. strategic nuclear deterrent force, or "triad," which also includes land-based intercontinental ballistic missiles (ICBMs) and land-based long-range bombers. At any given moment, some of the Navy's SSBNs are conducting nuclear deterrent patrols. The Navy's report on its FY2011 30-year shipbuilding plan states: "These ships are the most survivable leg of the Nation's strategic arsenal and provide the Nation's only day-to-day assured nuclear response capability."[7] The Department of Defense's (DOD's) report on the 2010 Nuclear Posture Review (NPR), released on April 6, 2010, states that "strategic nuclear submarines (SSBNs) and the SLBMs they carry represent the most survivable leg of the U.S. nuclear Triad."[8]

Current Ohio-Class SSBNs

The Navy currently operates 14 Ohio (SSBN-726) class SSBNs. The boats are commonly called Trident SSBNs or simply Tridents because they carry Trident SLBMs.

A total of 18 Ohio-class SSBNs were procured in FY1974-FY1991. The ships entered service in 1981-1997. The boats were designed and built by General Dynamics' Electric Boat Division (GD/EB) of Groton, CT, and Quonset Point, RI. They were originally designed for 30-year service lives but were later certified for 42-year service lives, consisting of two approximately 19-year periods of operation separated by an approximately four-year mid-life nuclear refueling overhaul, called an engineered refueling overhaul (ERO). The nuclear refueling overhaul includes both a nuclear refueling and overhaul work on the ship that is not related to the nuclear refueling.

(...continued)

nuclear initiative announced by President George H. W. Bush in September 1991. The initiative reserved a right to rearm SSNs at some point in the future with nuclear-armed Tomahawk land attack missiles (TLAM-Ns) should conditions warrant. Navy TLAM-Ns were placed in storage to support this option. DOD's report on the 2010 Nuclear Posture Review (NPR), released on April 6, 2010, states that the United States will retire the TLAM-Ns. (Department of Defense, *Nuclear Posture Review Report*, April 2010, pp. xiii and 28.)

[5] SSBNs, like other Navy submarines, are also equipped with horizontal torpedo tubes in the bow for firing torpedoes or other torpedo-sized weapons.

[6] This informal name is a reference to the large boom that would be made by the detonation of an SLBM nuclear warhead.

[7] U.S. Navy, *Report to Congress on Annual Long-Range Plan for Construction of Naval Vessels for FY 2011*, February 2010, p. 15.

[8] Department of Defense, *Nuclear Posture Review Report*, April 2010, p. 22. The next sentence in the report states: "Today, there appears to be no viable near or mid-term threats to the survivability of U.S. SSBNs, but such threats—or other technical problems—cannot be ruled out over the long term." The report similarly states on page 23: "Today, there appears to be no credible near or mid-term threats to the survivability of U.S. SSBNs. However, given the stakes involved, the Department of Defense will continue a robust SSBN Security Program that aims to anticipate potential threats and develop appropriate countermeasures to protect current and future SSBNs."

Ohio-class SSBNs are designed to each carry 24 SLBMs, although by 2018, four SLBM launch tubes on each boat are to be deactivated, and the number of SLBMs that can be carried by each boat consequently is to be reduced to 20, so that the number of operational launchers and warheads in the U.S. force will comply with strategic nuclear arms control limits.

The first eight boats in the class were originally armed with Trident I C-4 SLBMs; the final 10 were armed with larger and more-capable Trident II D-5 SLBMs. The Clinton Administration's 1994 Nuclear Posture Review (NPR) recommended a strategic nuclear force for the START II strategic nuclear arms reduction treaty that included 14 Ohio-class SSBNs, all armed with D-5s. This recommendation prompted interest in the idea of converting the first four Ohio-class boats (SSBNs 726-729) into SSGNs, so as to make good use of the 20 years of potential operational life remaining in these four boats, and to bolster the U.S. SSN fleet. The first four Ohio-class boats were converted into SSGNs in 2002-2008,[9] and the next four (SSBNs 730-733) were backfitted with D-5 SLBMs in 2000-2005, producing the current force of 14 Ohio-class SSBNs, all of which are armed with D-5 SLBMs.

Eight of the 14 Ohio-class SSBNs are homeported at Bangor, WA, in Puget Sound; the other six are homeported at Kings Bay, GA, close to the Florida border.

Unlike most Navy ships, which are operated by single crews, Navy SSBNs are operated by alternating crews (called the Blue and Gold crews) so as to maximize the percentage of time that they spend at sea in deployed status. The Navy consequently maintains 28 crews to operate its 14 Ohio-class SSBNs.

The first of the 14 Ohio-class SSBNs (SSBN-730) will reach the end of its 42-year service life in 2027. The remaining 13 will reach the ends of their service lives at a rate of roughly one ship per year thereafter, with the 14th reaching the end of its service life in 2040.

The Navy has initiated a program to refurbish and extend the service lives of D-5 SLBMs to 2042 "to match the OHIO Class submarine service life."[10]

Figure 1 shows an Ohio-class SSBN with the hatches to some of its SLBM launch tubes open.

---

[9] For more on the SSGN conversion program, see CRS Report RS21007, *Navy Trident Submarine Conversion (SSGN) Program: Background and Issues for Congress*, by Ronald O'Rourke.

[10] Statement of Rear Admiral Stephen Johnson, USN, Director, Strategic Systems Programs, Before the Subcommittee on Strategic Forces of the Senate Armed Services Committee [on] FY2011 Strategic Systems, March 17, 2010, p. 4.

Figure 1. Ohio (SSBN-726) Class SSBN

With the hatches to some of its SLBM launch tubes open

Source: U.S. Navy file photo accessed by CRS on February 24, 2011, at http://www.navy.mil/management/photodb/photos/101029-N-1325N-005.jpg.

Summary of U.S. SSBN Designs

The Navy has operated four classes of SSBNs since 1959. **Table 1** compares the current Ohio-class SSBN design to the three earlier U.S. SSBN designs. As shown in the table, the size of U.S. SSBNs has grown over time, reflecting in part a growth in the size and number of SLBMs carried on each boat. The Ohio class carries an SLBM (the D-5) that is much larger than the SLBMs carried by earlier U.S. SSBNs, and it carries 24 SLBMs, compared to the 16 on earlier U.S. SSBNs.[11] In part for these reasons, the Ohio-class design, with a submerged displacement of 18,750 tons, is more than twice the size of earlier U.S. SSBNs.

---

[11] The larger size of the Ohio-class design also reflects a growth in size over time in U.S. submarine designs due to other reasons, such as providing increased interior volume for measures to quiet the submarine acoustically, so as to make it harder to detect.

Table 1. U.S. SSBN Classes

	George Washington (SSBN-598) class	Ethan Allen (SSBN-608) class	Lafayette/Benjamin Franklin (SSBN-616/640) class	Ohio (SSBN-726) class
Number in class	5	5	31	18/14
Fiscal years procured	FY1958-FY1959	FY1959 and FY1961	FY1961-FY1964	FY1974/FY1977 - FY1991
Years in commission	1959-1985	1961-1992	1963-2002	1981/1984 - present
Length	381.7 feet	410.5 feet	425 feet	560 feet
Beam	33 feet	33 feet	33 feet	42 feet
Submerged displacement	6,700 tons	7,900 tons	8,250 tons	18,750 tons
Number of SLBM launch tubes	16	16	16	24 (to be reduced to 20 by 2018)
Final type(s) of SLBM carried	Polaris A-3	Polaris A-3	Poseidon C-3/ Trident I C-4	Trident II D-5
Diameter of those SLBMs	54 inches	54 inches	74 inches	83 inches
Length of those SLBMs	32.3 feet	32.3 feet	34 feet	44 feet
Weight of each SLBM (pounds)	36,000 pounds	36,000 pounds	65,000/73,000 pounds	~130,000 pounds
Range of SLBMs	~2,500 nm	~2,500 nm	~2,500 nm/~4,000 nm	~4,000 nm

Sources: Prepared by CRS based on data in Norman Polmar, *The Ships and Aircraft of the U.S. Fleet,* Annapolis, Naval Institute Press, various editions, and (for SSBN decommissioning dates) U.S. Naval Vessel Register.

Notes: Beam is the maximum width of a ship. For the submarines here, which have cylindrical hulls, beam is the diameter of the hull.

The range of an SLBM can vary, depending on the number and weight of nuclear warheads it carries; actual ranges can be lesser or greater than those shown.

The George Washington-class boats were procured as modifications of SSNs that were already under construction. Three of the boats were converted into SSNs toward the ends of their lives and were decommissioned in 1983-1985. The two boats that remained SSBNs throughout their lives were decommissioned in 1981.

All five Ethan Allen-class boats were converted into SSNs toward the ends of their lives. The boats were decommissioned in 1983 (two boats), 1985, 1991, and 1992.

Two of the Lafayette/Benjamin Franklin-class boats were converted into SSNs toward the ends of their lives and were decommissioned in 1999 and 2002. The 29 that remained SSBNs throughout their lives were decommissioned in 1986-1995. For 19 of the boats, the Poseidon C-3 was the final type of SLBM carried; for the other 12, the Trident I C-4 SLBM was the final type of SLBM carried.

A total of 18 Ohio-class SSBNs were built. The first four, which entered service in 1981-1984, were converted into SSGNs in 2002-2008. The remaining 14 boats entered service in 1984-1997. Although Ohio-class SSBNs are designed to each carry 24 SLBMs, by 2018, four SLBM launch tubes on each boat are to be deactivated, and the number of SLBMs that can be carried by each boat consequently is to be reduced to 20, so that the number of operational launchers and warheads in the U.S. force will comply with strategic nuclear arms control limits.

U.S.-UK Cooperation on SLBMs and the New UK SSBN

SSBNs are also operated by the United Kingdom, France, Russia, China, and India. The UK's four Vanguard-class SSBNs, which entered service in 1993-1999, each carry 16 Trident II D-5 SLBMs. Previous classes of UK SSBNs similarly carried earlier-generation U.S. SLBMs.[12] The UK's use of U.S.-made SLBMs on its SSBNs is one element of a long-standing close cooperation between the two countries on nuclear-related issues that is carried out under the 1958 Agreement for Cooperation on the Uses of Atomic Energy for Mutual Defense Purposes (also known as the Mutual Defense Agreement). Within the framework established by the 1958 agreement, cooperation on SLBMs in particular is carried out under the 1963 Polaris Sales Agreement and a 1982 Exchange of Letters between the two governments.[13] The Navy testified in March 2010 that "the United States and the United Kingdom have maintained a shared commitment to nuclear

[12] Although the SLBMs on UK SSBNs are U.S.-made, the nuclear warheads on the missiles are of UK design and manufacture.

[13] A March 18, 2010, report by the UK Parliament's House of Commons Foreign Affairs Committee stated:

During the Cold War, the UK's nuclear co-operation with the United States was considered to be at the heart of the [UK-U.S.] 'special relationship'. This included the 1958 Mutual Defence Agreement, the 1963 Polaris Sales Agreement (PSA) (subsequently amended for Trident), and the UK's use of the US nuclear test site in Nevada from 1962 to 1992. The co-operation also encompassed agreements for the United States to use bases in Britain, with the right to store nuclear weapons, and agreements for two bases in Yorkshire (Fylingdales and Menwith Hill) to be upgraded to support US missile defence plans.

In 1958, the UK and US signed the Mutual Defence Agreement (MDA). Although some of the appendices, amendments and Memoranda of Understanding remain classified, it is known that the agreement provides for extensive co-operation on nuclear warhead and reactor technologies, in particular the exchange of classified information concerning nuclear weapons to improve design, development and fabrication capability. The agreement also provides for the transfer of nuclear warhead-related materials. The agreement was renewed in 2004 for another ten years.

The other major UK-US agreement in this field is the 1963 Polaris Sales Agreement (PSA) which allows the UK to acquire, support and operate the US Trident missile system. Originally signed to allow the UK to acquire the Polaris Submarine Launched Ballistic Missile (SLBM) system in the 1960s, it was amended in 1980 to facilitate purchase of the Trident I (C4) missile and again in 1982 to authorise purchase of the more advanced Trident II (D5) in place of the C4. In return, the UK agreed to formally assign its nuclear forces to the defence of NATO, except in an extreme national emergency, under the terms of the 1962 Nassau Agreement reached between President John F. Kennedy and Prime Minister Harold Macmillan to facilitate negotiation of the PSA.

Current nuclear co-operation takes the form of leasing arrangements of around 60 Trident II D5 missiles from the US for the UK's independent deterrent, and long-standing collaboration on the design of the W76 nuclear warhead carried on UK missiles. In 2006 it was revealed that the US and the UK had been working jointly on a new 'Reliable Replacement Warhead' (RRW) that would modernise existing W76-style designs. In 2009 it emerged that simulation testing at Aldermaston on dual axis hydrodynamics experiments had provided the US with scientific data it did not otherwise possess on this RRW programme.

The level of co-operation between the two countries on highly sensitive military technology is, according to the written submission from Ian Kearns, "well above the norm, even for a close alliance relationship". He quoted Admiral William Crowe, the former US Ambassador to London, who likened the UK-US nuclear relationship to that of an iceberg, "with a small tip of it sticking out, but beneath the water there is quite a bit of everyday business that goes on between our two governments in a fashion that's unprecedented in the world." Dr Kearns also commented that the personal bonds between the US/UK scientific and technical establishments were deeply rooted.

(House of Commons, Foreign Affairs Committee, *Sixth Report Global Security: UK-US Relations*, March 18, 2010, paragraphs 131-135; http://www.publications.parliament.uk/pa/cm200910/cmselect/cmfaff/114/11402 htm; paragraphs 131-135 are included in the section of the report available at http://www.publications.parliament.uk/pa/cm200910/cmselect/cmfaff/114/11406 htm.)

deterrence through the Polaris Sales Agreement since April 1963. The U.S. will continue to maintain its strong strategic relationship with the UK for our respective follow-on platforms, based upon the Polaris Sales Agreement."[14]

The first Vanguard-class SSBN was originally projected to reach the end of its service life in 2024, but an October 2010 UK defense and security review report states that the lives of the Vanguard class ships will now be extended by a few years, so that the four boats will remain in service into the late 2020s and early 2030s.[15]

The UK plans to replace the four Vanguard-class boats with three or four next-generation SSBNs called Successor class SSBNs. The October 2010 UK defense and security review report states that each new Successor class SSBN is to be equipped with 8 D-5 SLBMs, rather than 12 as previously planned. The report states that "'Initial Gate'—a decision to move ahead with early stages of the work involved—will be approved and the next phase of the project will start by the end of [2010]. 'Main Gate'—the decision to start building the submarines—is required around 2016."[16] The first new boat is to be delivered by 2028, or about four years later than previously planned.[17]

The UK has wanted the Successor SSBNs to carry D-5 SLBMs, and for any successor to the D-5 SLBM to be compatible with, or be capable of being made compatible with, the D-5 launch system. President George W. Bush, in a December 2006 letter to UK Prime Minister Tony Blair, invited the UK to participate in any program to replace the D-5 SLBMs, and stated that any successor to the D-5 system should be compatible with, or be capable of being made compatible with, the launch system for the D-5 SLBM.

The United States is assisting the UK with certain aspects of the Successor SSBN program. In addition to the modular Common Missile Compartment (CMC) discussed below (see "Common Missile Compartment (CMC)" in the following section on the Ohio replacement program), the United States is assisting the UK with the new PWR-3 reactor plant[18] to be used by the Successor SSBN. A December 2011 press report states that "there has been strong [UK] collaboration with the US [on the Successor program], particularly with regard to the CMC, the PWR, and other propulsion technology," and that the design concept selected for the Successor class employs "a new propulsion plant based on a US design, but using next-generation UK reactor technology (PWR-3) and modern secondary propulsion systems."[19] The U.S. Navy states that

---

[14] Statement of Rear Admiral Stephen Johnson, USN, Director, Strategic Systems Programs, Before the Subcommittee on Strategic Forces of the Senate Armed Services Committee [on] FY2011 Strategic Systems, March 17, 2010, p. 6.

[15] *Securing Britain in an Age of Uncertainty: The Strategic Defence and Security Review*, Presented to Parliament by the Prime Minister by Command of Her Majesty, October 2010, p. 39.

[16] *Securing Britain in an Age of Uncertainty: The Strategic Defence and Security Review*, Presented to Parliament by the Prime Minister by Command of Her Majesty, October 2010, p. 5, 38-39. For more on the UK's Successor SSBN program as it existed prior to the October 2010 UK defense and security review report, see Richard Scott, "Deterrence At A Discount?" *Jane's Defence Weekly*, December 23, 2009: 26-31.

[17] *Securing Britain in an Age of Uncertainty: The Strategic Defence and Security Review*, Presented to Parliament by the Prime Minister by Command of Her Majesty, October 2010, p. 39.

[18] PWR3 means pressurized water reactor, design number 3. U.S. and UK nuclear-powered submarines employ pressurized water reactors. Earlier UK nuclear-powered submarines are powered by reactor designs that the UK designated PWR-2 and PWR-1.

[19] Sam LaGrone and Richard Scott, "Strategic Assets: Deterrent Plans Confront Cost Challenges," *Jane's Navy International*, December 2011: 17 and 18.

Naval Reactors, a joint Department of Energy/Department of Navy organization responsible for all aspects of naval nuclear propulsion, has an ongoing technical exchange with the UK Ministry of Defence under the US/UK 1958 Mutual Defence Agreement. The US/UK 1958 Mutual Defence Agreement is a Government to Government Atomic Energy Act agreement that allows the exchange of naval nuclear propulsion technology between the US and UK.

Under this agreement, Naval Reactors is providing the UK Ministry of Defence with US naval nuclear propulsion technology to facilitate development of the naval nuclear propulsion plant for the UK's next generation SUCCESSOR ballistic missile submarine. The technology exchange is managed and led by the US and UK Governments, with participation from Naval Reactors prime contractors, private nuclear capable shipbuilders, and several suppliers. A UK based office comprised of about 40 US personnel provide full-time engineering support for the exchange, with additional support from key US suppliers and other US based program personnel as needed.

The relationship between the US and UK under the 1958 mutual defence agreement is an ongoing relationship and the level of support varies depending on the nature of the support being provided. Naval Reactors work supporting the SUCCESSOR submarine is reimbursed by the UK Ministry of Defence.[20]

U.S. assistance to the UK on naval nuclear propulsion technology first occurred many years ago: To help jumpstart the UK's nuclear-powered submarine program, the United States transferred to the UK a complete nuclear propulsion plant (plus technical data, spares, and training) of the kind installed on the U.S. Navy's six Skipjack (SSN-585) class nuclear-powered attack submarines (SSNs), which entered service between 1959 and 1961. The plant was installed on the UK Navy's first nuclear-powered ship, the attack submarine *Dreadnought*, which entered service in 1963.

The December 2011 press report states that "the UK is also looking at other areas of cooperation between Successor and the Ohio Replacement Programme. For example, a collaboration agreement has been signed off regarding the platform integration of sonar arrays with the respective combat systems."[21]

Ohio Replacement Program

Program Origin and Early Milestones

Although the eventual need to replace the Ohio-class SSBNs has been known for many years, the Ohio replacement program can be traced more specifically to an exchange of letters in December 2006 between President George W. Bush and UK Prime Minister Tony Blair concerning the UK's desire to participate in a program to extend the service life of the Trident II D-5 SLBM into the 2040s, and to have its next-generation SSBNs carry D-5s. Following this exchange of letters, and with an awareness of the projected retirement dates of the Ohio-class SSBNs and the time that would likely be needed to develop and field a replacement for them, DOD in 2007 began studies on a next-generation sea-based strategic deterrent (SBSD).[22] The studies used the term sea-based

---

[20] Source: Email to CRS from Navy Office of Legislative Affairs, June 25, 2012.

[21] Sam LaGrone and Richard Scott, "Strategic Assets: Deterrent Plans Confront Cost Challenges," *Jane's Navy International*, December 2011: 19.

[22] In February 2007, the commander of U.S. Strategic Command (STRATCOM) commissioned a task force to support an anticipated Underwater Launched Missile Study (ULMS). On June 8, 2007, the Secretary of the Navy initiated the (continued...)

strategic deterrent (SBSD) to signal the possibility that the new system would not necessarily be a submarine.

An Initial Capabilities Document (ICD) for a new SBSD was developed in early 2008[23] and approved by DOD's Joint Requirements Oversight Committee (JROC) on June 20, 2008.[24] In July 2008, DOD issued a Concept Decision providing guidance for an analysis of alternatives (AOA) for the program; an acquisition decision memorandum from John Young, DOD's acquisition executive, stated the new system would, barring some discovery, be a submarine.[25] The Navy established an Ohio replacement program office at about this same time.[26]

The AOA reportedly began in the summer or fall of 2008.[27] The AOA was completed, with final brief to the Office of the Secretary of Defense (OSD), on May 20, 2009. The final AOA report was completed in September 2009. An AOA Sufficiency Review Letter was signed by OSD's Director, Cost Assessment & Program Evaluation (CAPE) on December 8, 2009.[28] The AOA concluded that a new-design SSBN was the best option for replacing the Ohio-class SSBNs. (For a June 26, 2013, Navy blog post discussing options that were examined for replacing the Ohio-class SSBNs, see **Appendix A**.)

The program's Milestone A review meeting was held on December 9, 2010. On February 3, 2011, the Navy provided the following statement to CRS concerning the outcome of the December 9 meeting:

> The OHIO Replacement Program achieved Milestone A and has been approved to enter the Technology Development Phase of the Dept. of Defense Life Cycle Management System as of Jan. 10, 2011.
>
> This milestone comes following the endorsement of the Defense Acquisition Board (DAB), chaired by Dr. Carter (USD for Acquisition, Technology, and Logistics) who has signed the program's Milestone A Acquisition Decision Memorandum (ADM).
>
> The DAB endorsed replacing the current 14 Ohio-class Ballistic Missile Submarines (SSBNs) as they reach the end of their service life with 12 Ohio Replacement Submarines,

---

(...continued)

ULMS. Six days later, the commander of STRATCOM directed that a Sea Based Strategic Deterrent (SBSD) capability-based assessment (CBA) be performed. In July 2007, the task force established by the commander of STRATCOM provided its recommendations regarding capabilities and characteristics for a new SBSD. (Source: Navy list of key events relating to the ULMS and SBSD provided to CRS and the Congressional Budget Office (CBO) on July 7, 2008.)

[23] On February 14, 2008, the SBSD ICD was approved for joint staffing by the Navy's Resources and Requirements Review Board (R3B). On April 29, 2008, the SBSD was approved by DOD's Functional Capabilities Board (FCB) to proceed to DOD's Joint Capabilities Board (JCB). (Source: Navy list of key events relating to the ULMS and SBSD provided to CRS and CBO on July 7, 2008.)

[24] Navy briefing to CRS and CBO on the SBSD program, July 6, 2009.

[25] Navy briefing to CRS and CBO on the SBSD program, July 6, 2009.

[26] An August 2008 press report states that the program office, called PMS-397, "was established within the last two months." (Dan Taylor, "Navy Stands Up Program Office To Manage Next-Generation SSBN," *Inside the Navy*, August 17, 2008.

[27] "Going Ballistic," *Defense Daily*, September 22, 2008, p. 1.

[28] *Department of Defense Fiscal Year (FY) 2012 Budget Estimates, Navy, Justification Book Volume 2, Research, Development, Test & Evaluation, Navy Budget Activity 4*, entry for PE0603561N, Project 3220 (pdf page 345 of 888).

each comprising 16, 87-inch diameter missile tubes utilizing TRIDENT II D5 Life Extended missiles (initial loadout). The decision came after the program was presented to the Defense Acquisition Board (DAB) on Dec. 9, 2010.

The ADM validates the program's Technology Development Strategy and allows entry into the Technology Development Phase during which warfighting requirements will be refined to meet operational and affordability goals. Design, prototyping, and technology development efforts will continue to ensure sufficient technological maturity for lead ship procurement in 2019.[29]

Planned Procurement Quantity: 12 SSBN(X)s to Replace 14 Ohio-Class Boats

Navy plans call for procuring 12 SSBN(X)s to replace the current force of 14 Ohio-class SSBNs. In explaining the planned procurement quantity of 12 boats, the Navy states that 10 operational SSBNs—meaning boats not encumbered by lengthy maintenance actions—are needed to meet strategic nuclear deterrence requirements for having a certain number of SSBNs at sea at any given moment. The Navy states that a force of 14 Ohio-class boats was needed to meet this requirement because, during the middle years of the Ohio class life cycle, three and sometimes four of the boats are non-operational at any given moment on account of being in the midst of lengthy mid-life nuclear refueling overhauls or other extended maintenance actions. The Navy states that 12 rather than 14 SSBN(X)s will be needed to meet the requirement for 10 operational boats because the mid-life overhauls of SSBN(X)s, which will not include a nuclear refueling, will require less time (about two years) than the mid-life refueling overhauls of Ohio-class boats (which require about four years from contract award to delivery),[30] the result being that only two SSBN(X)s (rather than three or sometimes four) will be in the midst of mid-life overhauls or other extended maintenance actions at any given moment during the middle years of the SSBN(X) class life cycle.[31]

Procurement and Replacement Schedule

Table 2 shows the Navy's proposed schedule for procuring 12 SSBN(X)s, and for having SSBN(X)s replace Ohio-class SSBNs. As shown in **Table 2**, under the Navy's FY2012 budget, the first Ohio replacement boat was scheduled to be procured in FY2019, and Ohio replacement boats were to enter service on a schedule that would maintain the Navy's SSBN force at 12 boats. As also shown in **Table 2**, the Navy's FY2013 budget deferred the procurement of the first Ohio replacement boat by two years, to FY2021. As a result of the deferment of the procurement of the lead boat from FY2019 to FY2021, the Navy's SSBN force will drop to 11 or 10 boats for the period FY2029-FY2041. The Navy states that the reduction to 11 or 10 boats during this period is acceptable in terms of meeting strategic nuclear deterrence requirements, because during these years, all 11 or 10 of the SSBNs in service will be operational (i.e., none of them will be in the

---

[29] Source: Email from Navy Office of Legislative Affairs to CRS, February 3, 2011.

[30] Navy budget submissions show that Ohio-class mid-life nuclear refueling overhauls have contract-award-to-delivery periods generally ranging from 47 months to 50 months.

[31] Source: Navy update briefing on Ohio replacement program to CRS and Congressional Budget Office (CBO), September 17, 2012. See also "Navy Responds to Debate Over the Size of the SSBN Force," Navy Live, May 16, 2013, accessed July 26, 2013, at http://navylive.dodlive mil/2013/05/16/navy-responds-to-debate-over-the-size-of-the-ssbn-force/, and Richard Breckenridge, "SSBN Force Level Requirements: It's Simply a Matter of Geography," Navy Live, July 19, 2013, accessed July 26, 2013, at http://navylive.dodlive mil/2013/07/19/ssbn-force-level-requirements-its-simply-a-matter-of-geography/.

midst of a lengthy mid-life overhaul). The Navy acknowledges that there is some risk in having the SSBN force drop to 11 or 10 boats, because it provides little margin for absorbing an unforeseen event that might force an SSBN into an unscheduled and lengthy maintenance action.[32] (See also the discussion above in "Planned Procurement Quantity: 12 SSBN(X)s to Replace 14 Ohio-Class Boats.")

The minimum level of 10 boats shown in **Table 2** for the period FY2032-FY2040 can be increased to 11 boats (providing some margin for absorbing an unforeseen event that might force an SSBN into an unscheduled and lengthy maintenance action) by accelerating by about one year the planned procurement dates of boats 2 through 12 in the program. Under this option, the second boat in the program would be procured in FY2023 rather than FY2024, the third boat in the program would be procured in FY2025 rather than FY2026, and so on. Implementing this option could affect the Navy's plan for funding the procurement of Virginia-class attack submarines during the period FY2022-FY2025.[33]

---

[32] Source: Navy update briefing on Ohio replacement program to CRS and Congressional Budget Office (CBO), September 17, 2012. A September 28, 2012, press report similarly quotes Rear Admiral Barry Bruner, the Navy's director of undersea warfare, as stating that "During this time frame, no major SSBN overhauls are planned, and a force of 10 SSBNs will support current at-sea presence requirements," and that "This provides a low margin to compensate for unforeseen issues that may result in reduced SSBN availability. The reduced SSBN availability during this time frame reinforces the importance of remaining on schedule with the Ohio Replacement program to meet future strategic requirements. As the Ohio Replacement ships begin their mid-life overhauls in 2049, 12 SSBNs will be required to offset ships conducting planned maintenance." (Michael Fabey, U.S. Navy Defends Boomer Submarine Replacement Plans," *Aerospace Daily & Defense Report*, September 28, 2012: 3.)

[33] For more on the Virginia-class program, see CRS Report RL32418, *Navy Virginia (SSN-774) Class Attack Submarine Procurement: Background and Issues for Congress*, by Ronald O'Rourke.

Table 2. Navy Schedule for Procuring SSBN(X)s and Replacing Ohio-Class SSBNs

Fiscal Year	Schedule in FY2012 Budget				Schedule Under Subsequent Budgets			
	Number of SSBN(X)s procured each year	Cumulative number of SSBN(X)s in service	Ohio-class SSBNs in service	Combined number of Ohio-class SSBNs and SSBN(X)s in service	Number of SSBN(X)s procured each year	Cumulative number of SSBN(X)s in service	Ohio-class SSBNs in service	Combined number of Ohio-class SSBNs and SSBN(X)s in service
2019	1		14	14			14	14
2020			14	14			14	14
2021			14	14	1		14	14
2022	1		14	14			14	14
2023			14	14			14	14
2024	1		14	14	1		14	14
2025	1		14	14			14	14
2026	1		14	14	1		14	14
2027	1		13	13	1		13	13
2028	1		12	13	1		12	12
2029	1	1	11	12	1		11	11
2030	1	2	10	12	1	1	10	11
2031	1	3	9	12	1	2	9	11
2032	1	4	8	12	1	2	8	10
2033	1	5	7	12	1	3	7	10
2034		6	6	12	1	4	6	10
2035		7	5	12	1	5	5	10
2036		8	4	12		6	4	10
2037		9	3	12		7	3	10
2038		10	2	12		8	2	10
2039		11	1	12		9	1	10
2040		12		12		10	0	10
2041		12		12		11	0	11
2042		12		12		12	0	12

Source: Navy FY2012-FY2015 budget submissions.

SSBN(X) Design Features

The design of the SSBN(X), now being developed, will reflect the following:

- The SSBN(X) is to be designed for a 40- or 42-year expected service life.[34]

- Unlike the Ohio-class design, which requires a mid-life nuclear refueling,[35] the SSBN(X) is to be equipped with a life-of-the-ship nuclear fuel core (a nuclear

---

[34] U.S. Navy, *Report to Congress on Annual Long-Range Plan for Construction of Naval Vessels for FY 2011*, February 2010, p. 24; and Sam LaGrone, "Navy Has Finalized Specifications for New Ohio-replacement Boomer," *USNI News* (http://news.usni.org), April 7, 2014.

[35] As mentioned earlier (see "Current Ohio-Class SSBNs"), the Ohio-class boats receive a mid-life nuclear refueling (continued...)

fuel core that is sufficient to power the ship for its entire expected service life).[36] Although the SSBN(X) will not need a mid-life nuclear refueling, it will still need a mid-life non-refueling overhaul (i.e., an overhaul that does not include a nuclear refueling) to operate over its full 40-year life.

- The SSBN(X) is to be equipped with an electric-drive propulsion train, as opposed to the mechanical-drive propulsion train used on other Navy submarines. The electric-drive system is expected to be quieter (i.e., stealthier) than a mechanical-drive system.[37]

- The SSBN(X) is to have SLBM launch tubes that are the same size as those on the Ohio class (i.e., tubes with a diameter of 87 inches and a length sufficient to accommodate a D-5 SLBM).

- The SSBN(X) will have a beam (i.e., diameter)[38] of 43 feet, compared to 42 feet on the Ohio-class design,[39] and a length of 560 feet, the same as that of the Ohio-class design.[40]

- Instead of 24 SLBM launch tubes, as on the Ohio-class design, the SSBN(X) is to have 16 SLBM launch tubes. (For further discussion of the decision to equip the boat with 16 tubes rather than 20, see **Appendix B**.)

- Although the SSBN(X) is to have fewer launch tubes than the Ohio-class SSBN, it is to be larger than the Ohio-class SSBN design, with a reported submerged displacement of more than 20,000 tons, compared to 18,750 tons for the Ohio-class design.[41]

- The Navy states that "owing to the unique demands of strategic relevance, [SSBN(X)s] must be fitted with the most up-to-date capabilities and stealth to ensure they are survivable throughout their full 40-year life span."[42]

(...continued)
overhaul, called an Engineered Refueling Overhaul (ERO), which includes both a nuclear refueling and overhaul work on the ship that is not related to the nuclear refueling.

[36] U.S. Navy, *Report to Congress on Annual Long-Range Plan for Construction of Naval Vessels for FY 2011*, February 2010, p. 5. The two most recent classes of SSNs—the Seawolf (SSN-21) and Virginia (SSN-774) class boats—are built with cores that are expected to be sufficient for their entire 33-year expected service lives.

[37] Source: Spoken testimony of Admiral Kirkland Donald, Deputy Administrator for Naval Reactors, and Director, Naval Nuclear Propulsion, National Nuclear Security Administration, at a March 30, 2011, hearing before the Strategic Forces Subcommittee of the Senate Armed Services Committee, as shown in the transcript of the hearing. See also Dave Bishop, "What Will Follow the Ohio Class?" *U.S. Naval Institute Proceedings*, June 2012: 31; and Sam LaGrone and Richard Scott, "Strategic Assets: Deterrent Plans Confront Cost Challenges," *Jane's Navy International*, December 2011: 16.

[38] Beam is the maximum width of a ship. For Navy submarines, which have cylindrical hulls, beam is the diameter of the hull.

[39] Dave Bishop, "What Will Follow the Ohio Class?" *U.S. Naval Institute Proceedings*, June 2012: 31. (Bishop was program manager for the Ohio replacement program.) See also Sam LaGrone and Richard Scott, "Strategic Assets: Deterrent Plans Confront Cost Challenges," *Jane's Navy International*, December 2011: 15 and 16.

[40] Sydney J. Freedberg, "Navy Seeks Sub Replacement Savings: From NASA Rocket Boosters To Reused Access Doors," *Breaking Defense* (http://breakingdefense.com), April 7, 2014.

[41] Sam LaGrone, "Navy Has Finalized Specifications for New Ohio-replacement Boomer," *USNI News* (http://news.usni.org), April 7, 2014.

[42] U.S. Navy, *Report to Congress on Annual Long-Range Plan for Construction of Naval Vessels for FY 2011*, February 2010, p. 24.

In an article published in June 2012, the program manager for the Ohio replacement program stated that "the current configuration of the Ohio replacement is an SSBN with 16 87-inch-diameter missile tubes, a 43-foot-diamater hull, fairwater planes,[43] electric-drive propulsion, [an] X-stern,[44] accommodations for 155 personnel, and a common submarine radio room[45] tailored to the SSBN mission."[46]

Acquisition Cost

A March 2014 GAO report assessing selected major DOD weapon acquisition programs states that the estimated total acquisition cost of the SSBN(X) program is $95,103.2 million (about $95.1 billion) in constant FY2014 dollars, including $11,718.2 million (about $11.7 billion) in research and development costs and $83,385.0 million (about $83.4 billion) in procurement costs.[47]

The Navy's FY2014 30-year shipbuilding plan, submitted in May 2013, estimates the procurement cost of the lead boat in the program at $12.0 billion in constant 2013 dollars, including $4.6 billion in detailed design and nonrecurring engineering (DD/NRE) costs for the entire class, and $7.4 billion in construction costs for the ship itself.[48] (It is a traditional budgeting practice for Navy shipbuilding programs to attach the DD/NRE costs for a new class of ships to the procurement cost of the lead ship in the class.)

The Navy in February 2010 preliminarily estimated the procurement cost of each Ohio replacement boat at $6 billion to $7 billion in FY2010 dollars.[49] Following the Ohio replacement program's December 9, 2010, Milestone A acquisition review meeting (see "Program Origin and Early Milestones"), DOD issued an Acquisition Decision Memorandum (ADM) that, among other things, established a target average unit procurement cost for boats 2-12 in the program of $4.9 billion in constant FY2010 dollars.[50] The Navy is working to achieve this target cost. In 2011, the Navy estimated that its cost-reduction efforts had reduced the estimated average unit

---

[43] The term fairwater planes means that the submarine's forward diving planes are mounted on the ship's hull, near the bow, rather than on the ship's sail (aka "conning tower").

[44] The term X-stern means that the steering and diving fins at the stern of the ship are, when viewed from the rear, in the diagonal pattern of the letter X, rather than the vertical-and horizontal pattern of a plus sign (which is referred to as a cruciform stern).

[45] The common submarine radio room is a standardized (i.e., common) suite of submarine radio room equipment that is being installed on other U.S. Navy submarines.

[46] Dave Bishop, "What Will Follow the Ohio Class?" *U.S. Naval Institute Proceedings*, June 2012: 31. See also Sam LaGrone and Richard Scott, "Strategic Assets: Deterrent Plans Confront Cost Challenges," *Jane's Navy International*, December 2011: 15 and 16.

[47] Government Accountability Office, *Defense Acquisitions[:] Assessments of Selected Weapons Programs*, GAO-14-340SP, March 2014, p. 141.

[48] Department of the Navy, *Report to Congress on the Long-Range Plan for Construction of Naval Vessels for FY2014*, May 2013, p. 15.

[49] Department of the Navy, *Report to Congress on Annual Long-Range Plan for Construction of Naval Vessels for FY 2011*, February 2010, p. 20.

[50] Christopher J. Castelli, "DOD: New Nuclear Subs Will Cost $347 Billion To Acquire, Operate," *Inside the Navy*, February 21, 2011; Elaine M. Grossman, "Future U.S. Nuclear-Armed Vessel to Use Attack-Submarine Technology," *Global Security Newswire*, February 24, 2011; Jason Sherman, "Navy Working To Cut $7.7 Billion From Ohio Replacement Program," *Inside the Navy*, February 28, 2011. See also Christopher J. Castelli, "DOD Puts 'Should-Cost' Pressure On Major Weapons Programs," *Inside the Navy*, May 2, 2011.

---

procurement cost of boats 2-12 to $5.6 billion each in constant FY2010 dollars.[51] In May 2013, the Navy stated that its continued cost-reduction efforts had reduced the estimated average unit procurement cost of boats 2-12 to about $5.4 billion each in constant FY2010 dollars.[52] In April 2014, the Navy stated that its continued cost-reduction efforts had reduced the estimated average unit procurement cost of boats 2-12 to about $5.36 billion each in constant FY2010 dollars.[53] The Navy continues examining potential further measures to bring the cost of boats 2-12 closer to the $4.9 billion target cost.[54]

The above cost figures do not include costs for refurbishing D-5 SLBMs so as to extend their service lives to 2042.

Operation and Support (O&S) Cost

The Navy is working to reduce the estimated operation and support (O&S) cost of each SSBN(X) from $124 million per year to $110 million per year in constant FY2010 dollars.[55]

Common Missile Compartment (CMC)

Current U.S. and UK plans call for the SSBN(X) and the UK's Successor SSBN to use a missile compartment—the middle section of the boat with the SLBM launch tubes—of the same general design.[56] As mentioned earlier (see "U.S.-UK Cooperation on SLBMs"), the UK's SSBN is to be armed with eight SLBMs, or half the number to be carried by the SSBN(X). The modular design of the CMC will accommodate this difference. Since the UK's first Vanguard-class SSBN was originally projected to reach the end of its service life in 2024—three years before the first Ohio-class SSBN is projected to reach the end of its service life—design work on the CMC began about three years sooner than would have been required to support the Ohio replacement program alone. This is the principal reason why the FY2010 budget included a substantial amount of research and development funding for the CMC. The UK is providing some of the funding for the design of the CMC, including a large portion of the initial funding.

A March 2010 Government Accountability office (GAO) report stated:

> According to the Navy, in February 2008, the United States and United Kingdom began a joint effort to design a common missile compartment. This effort includes the participation of government officials from both countries, as well as industry officials from Electric Boat

---

[51] Source: Navy briefing for CRS and Congressional Budget Office on Navy submarine programs, March 16, 2012.

[52] Source: Navy meeting with CRS and CBO to discuss Navy acquisition issues, May 17, 2013.

[53] Sam LaGrone, "Navy Has Finalized Specifications for New Ohio-replacement Boomer," *USNI News* (http://news.usni.org), April 7, 2014; Sydney J. Freedberg, "Navy Seeks Sub Replacement Savings: From NASA Rocket Boosters To Reused Access Doors," *Breaking Defense* (http://breakingdefense.com), April 7, 2014.

[54] See, for example, Dave Bishop, "Two Years In And Ground Strong, The Ohio Replacement Program," *Undersea Warfare*, Spring 2012: 5-7.

[55] Dave Bishop, "Two Years In And Ground Strong, The Ohio Replacement Program," *Undersea Warfare*, Spring 2012: 5; Megan Eckstein, "Ohio-Replacement Sub Technology To Drop O&S Costs To $110M A Year," *Inside the Navy*, March 1, 2013.

[56] Statement of Rear Admiral Stephen Johnson, USN, Director, Strategic Systems Programs, Before the Subcommittee on Strategic Forces of the Senate Armed Services Committee [on] FY2011 Strategic Systems, March 17, 2010, p. 6, which states: "The OHIO Replacement programs includes the development of a common missile compartment that will support both the OHIO Class Replacement and the successor to the UK Vanguard Class."

Corporation and BAE Systems. To date, the United Kingdom has provided a larger share of funding for this effort, totaling just over $200 million in fiscal years 2008 and 2009.[57]

A March 2011 GAO report stated:

> The main focus of OR [Ohio Replacement program] research and development to date has been the CMC. The United Kingdom has provided $329 million for this effort since fiscal year 2008. During fiscal years 2009 and 2010, the Navy had allocated about $183 million for the design and prototyping of the missile compartment.[58]

A May 2010 press report stated that "the UK has, to date, funded the vast majority of [the CMC's] upfront engineering design activity and has established a significant presence in Electric Boat's Shaw's Cove CMC design office in New London, CT."[59]

Under the October 2010 UK defense and security review report (see "U.S.-UK Cooperation on SLBMs"), the UK now plans to deliver its first Successor class SSBN in 2028, or about four years later than previously planned.

Program Funding

Table 3 shows funding for the Ohio replacement program. The table shows U.S. funding only; it does not include funding provided by the UK to help pay for the design of the CMC. As can be seen in the table, the Navy's proposed FY2015 budget requests $1,219.3 million for continued research and development work on the program.

Table 3. Ohio Replacement Program Funding

(Millions of then-year dollars, rounded to nearest tenth; totals may not add due to rounding)

	FY08	FY09	FY10	FY11	FY12	FY13	FY14	FY15 (req.)	FY16 (proj.)	FY17 (proj.)	FY18 (proj.	FY19 (proj.)
PE0101221N/Project 3198	0	9.7	0	0	0	0	0	0	0	0	0	0
PE0603561N/Project 3220	0	0	363.4	431.4	761.2	431.9	784.8	0	0	0	0	0
PE0603561N/Project 9999	4.9	3.2	4.0	0	0	0	0	0	0	0	0	0
PE0603570N/Project 3219	0	0	107.5	178.3	285.4	73.7	296.1	370.0	422.7	411.6	401.7	291.3
PE0603595N/Project 3220	0	0	0	0	0	0	0	812.8	994.9	696.3	709.5	394.5
PE0603595N/Project 3237	0	0	0	0	0	0	0	36.5	0	0	0	0
Total R&D funding	**4.9**	**12.9**	**474.9**	**609.7**	**1,046.6**	**505.6**	**1,080.9**	**1,219.3**	**1,417.6**	**1,107.9**	**1,111.2**	**685.8**
Procurement funding	**0**	**0**	**0**	**0**	**0**	**0**	**0**	**0**	**13.2**	**777.8**	**791.8**	**2,887.9**
TOTAL all funding	**4.9**	**12.9**	**474.9**	**609.7**	**1.046.6**	**505.6**	**1,080.9**	**1,219.3**	**1,430.8**	**1,885.7**	**1,903.0**	**3,573.7**

---

[57] Government Accountability Office, *Defense Acquisitions[:] Assessments of Selected Weapon Programs*, GAO-10-388SP, March 2010, p. 152.

[58] Government Accountability Office, *Defense Acquisitions[:] Assessments of Selected Weapon Programs*, GAO-11-233SP, March 2011, p. 147.

[59] Sam LaGrone and Richard Scott, "Deterrent Decisions: US and UK Wait on Next Steps for SSBN Replacements," *Jane's Navy International*, May 2010, pp. 10-11.

Source: Navy FY2015 budget and prior-year budgets.

Notes: PE means Program Element, that is, a research and development line item. A Program Element may include several projects. **PE0101221N/Project 3198** is Underwater Launch Missile System (ULMS) project within the PE for Strategic Submarine and Weapons System Support. **PE0603561N/Project 3220** is Sea-Based Strategic Deterrent (SBSD) project within the PE for Advanced Submarine System Development. **PE0603561N/Project 9999** is Congressional funding additions within the PE for Advanced Submarine System Development. **PE0603570N/Project 3219** is SSBN(X) reactor plant project within the PE for Advanced Nuclear Power Systems. **PE0603595N/Project 3220** is Sea-Based Strategic Deterrent (SBSD) Advanced Submarine System Development project within the PE for Ohio replacement. **PE0603595N/Project 3237** is Launch Test Facility project within the PE for Ohio Replacement. **Procurement funding** shown in FY2017 and FY2018 is advance procurement funding for first SSBN(X), which is scheduled to be procured in FY2021.

Issues for Congress

Funding Shortfall for Manufacturing Nuclear Fuel Core

One potential oversight issue for Congress concerns funding for manufacturing the nuclear fuel core for the lead ship in the program. A March 27, 2014, press report states:

> The Navy's "highest-priority program," the Ohio-class Replacement, could face a six month delay due to a funding shortfall in fiscal 2014 for the manufacture of the submarine's reactor core.
>
> Chief of Naval Operations ADM Jonathan W. Greenert made the announcement during a March 27 Senate Armed Services Committee hearing. "We need to reconcile this. ... We will reconcile this," he said.
>
> The possible delay focuses on a $150 million gap, and Greenert said he was going to work with ADM John Richardson, director of the Naval Nuclear Propulsion Program, on a solution.
>
> In an interview with Seapower, Richardson, who was not part of the hearing panel, said the funding hole was coming from the Department of Energy (DOE). He said if given the additional funds in fiscal 2015, he can make up the six-months. If not, it would further delay the program.
>
> "The schedule for the Ohio-class replacement, including the design and transition to construction, is very aggressive. There's a lot of parts that have to move together to make that happen on time," he said.
>
> Richardson's team is designing and overseeing the construction of the submarine's reactor core. The shortage of funding in fiscal 2014—which was marked against his technological support base—did not allow him to purchase a high-preforming computer that is part of that design.
>
> The admiral stressed that the program currently is on track, but he cannot recover the time lost if he's not given the appropriate funding in fiscal 2015.

He said he could not speculate on how long additional delays would last if he's not given the needed funding because he's not done that analysis yet. "Even with reprioritization [of funds] it would become impossible to make up the time," he said.[60]

Likelihood That Navy Will Reach $4.9 Billion Target Cost

Another potential oversight issue for Congress regarding the Ohio replacement program is the likelihood that the Navy will be able to achieve DOD's goal of reducing the average unit procurement cost of boats 2-12 in the program to $4.9 billion each in FY2010 dollars. As mentioned earlier, as of April 2014, the Navy estimated that its cost-reduction efforts had reduced the average unit procurement cost of boats 2-12 to about $5.36 billion each in FY2010 dollars, leaving another $460 million or so in cost reduction to reach the $4.9 billion target cost. Measures that the Navy has taken to reduce the average unit procurement cost of boats 2-12 to about $5.36 billion include, among other things, reducing the number of SLBM launch tubes from 20 to 16,[61] and making the launch tubes no larger in diameter than those on the Ohio-class design.[62]

An October 19, 2012, press report quoted Rear Admiral David Johnson, the program executive officer for submarines, as stating that in achieving the targeted reduction in per-boat procurement cost, "I think one of the biggest effectors we can do is buying the ship smartly.... We can probably get somewhere in the range of $300 million-plus per ship out [of the estimated cost], just by buying the ships smartly, encouraging a long production run in industry and the vendor base."[63] An April 7, 2014, press report stated:

> "We are looking at everything," [Rear Admiral David] Johnson told reporters, "all the way down to trying to reuse the doors on the missile tube access covers from the Ohio" as those subs go out of service. "Those doors are dry"—i.e. they aren't exposed to the ocean—"so they really see no wear," he said.
>
> It's relatively easy to reuse missile tube parts because the tubes themselves are the same size on both the Ohios and the future missile sub, which will also carry the Trident for at least the first part of its service life. (An all-new nuclear missile is a notion for the distant future). But nobody's building Ohios any more, so Johnson's priority is taking advantage of the Navy's ongoing Virginia--class attack sub program.
>
> The service is steadily buying two Virginia submarines a year to add to the 10 already in service. By contrast, the entire Ohio Replacement Program (formerly known as SSN(X)) will be 12 subs, so any way to piggyback off the higher-volume program will save money. Johnson wants to bundle procurement of at least some materials and components that will go on both submarines.
>
> So how many components will the Virginia and the ORP have in common? There's not even an estimate yet, Johnson said. "It's not like ten percent, it's not like 75%, it's somewhere in

---

[60] John C. Marcario, "Lack of DOE Funds Could Delay Ohio-class Replacement Program," *Seapower (www.seapowermagazine.org)*, March 27, 2014.

[61] For further discussion of the decision to equip the boat with 16 tubes rather than 20, see **Appendix B**.

[62] The Navy had examined the option of equipping the SSBN(X) with tubes greater in diameter than those on the Ohio-class design, so as to support an option of arming the boats many years from now with a new SLBM that is larger in diameter than the D-5 SLBM.

[63] As quoted in Lee Hudson, "Navy Plans To Award Ohio Replacement R&D Contract In December," *Inside the Navy*, October 22, 2012.

the range there," he said. But as the Navy and industry design the replacement for the Ohio-class, he said, with every component, "we see if we can make it fit using a Seawolf or Ohio or Virginia-class pump, valve," etc.

But since the new nuclear missile submarine will be larger than anything now in service—the biggest submarine ever built in the US, said Johnson, roughly twice the size of the Virginia—a lot of its components will have to be bigger, too. Even with those scaled-up parts, though, the admiral said, the same factory can often build a big version and a little version of a given component, a pump for example, at a lower cost than two companies building entirely different designs.[64]

Potential oversight questions include the following:

- How did DOD settle on the figure of $4.9 billion in FY2010 dollars as the target average unit procurement cost for boats 2-12 in the program? On what analysis was the selection of this figure based?

- How difficult will it be for the Navy to reach this target cost? What options is the Navy examining to achieve the additional $460 million or so in unit procurement cost savings needed to reach it?

- Would a boat costing $4.9 billion have sufficient capability to perform its intended missions?

- What, if anything, does DOD plan to do if the Navy is unable to achieve the $4.9 billion target cost figure? If $4.9 billion is the target figure, is there a corresponding "ceiling" figure higher than $4.9 billion, above which DOD would not permit the Ohio replacement program to proceed? If no such figure exists, should DOD establish one?

Accuracy of Navy's Estimated Unit Procurement Cost

Overview

Another potential oversight issue for Congress concerns the accuracy of the Navy's estimate of the procurement cost of each SSBN(X). The accuracy of the Navy's estimate is a key consideration in assessing the potential affordability of the Ohio replacement program, including its potential impact on the Navy's ability to procure other kinds of ships during the years of SSBN(X) procurement. Some of the Navy's ship designs in recent years, such as the Gerald R. Ford (CVN-78) class aircraft carrier,[65] the San Antonio (LPD-17) class amphibious ship[66] and the Littoral Combat Ship (LCS),[67] have proven to be substantially more expensive to build than the Navy originally estimated.

---

[64] Sydney J. Freedberg, "Navy Seeks Sub Replacement Savings: From NASA Rocket Boosters To Reused Access Doors," *Breaking Defense* (http://breakingdefense.com), April 7, 2014.

[65] For more on the CVN-78 program, see CRS Report RS20643, *Navy Ford (CVN-78) Class Aircraft Carrier Program: Background and Issues for Congress*, by Ronald O'Rourke.

[66] For more on the LPD-17 program, see CRS Report RL34476, *Navy LPD-17 Amphibious Ship Procurement: Background, Issues, and Options for Congress*, by Ronald O'Rourke.

[67] For more on the LCS program, see CRS Report RL33741, *Navy Littoral Combat Ship (LCS) Program: Background and Issues for Congress*, by Ronald O'Rourke.

The accuracy of the Navy's estimate can be assessed in part by examining known procurement costs for other recent Navy submarines—including Virginia (SSN-774) class attack submarines (which are currently being procured), Seawolf (SSN-21) class attack submarines (which were procured prior to the Virginia class), and Ohio (SSBN-726) class ballistic missile submarines— and then adjusting these costs for the Ohio replacement program so as to account for factors such as differences in ship displacement and design features, changes over time in submarine technologies (which can either increase or reduce a ship's procurement cost, depending on the exact technologies in question), advances in design for producibility (i.e., design features that are intended to make ships easier to build), advances in shipyard production processes (such as modular construction), and changes in submarine production economies of scale (i.e., changes in the total number of attack submarines and ballistic missile submarines under construction at any one time).

The Navy's estimated unit procurement cost for the program at any given point will reflect assumptions on, among other things, which shipyard or shipyards will build the boats, and how much Virginia-class construction will be taking place in the years when SSBN(X)s are being built. Changing the Navy's assumption about which shipyard or shipyards will build SSBN(X)s could reduce or increase the Navy's estimated unit procurement cost for the boats. If shipbuilding affordability pressures result in Virginia-class boats being removed from the 30-year shipbuilding plan during the years of SSBN(X) procurement, the resulting reduction in submarine production economies of scale could make SSBN(X)s more expensive to build than the Navy estimates.

October 2013 CBO Report

An October 2013 Congressional Budget Office (CBO) report on the cost of the Navy's shipbuilding programs stated:

> The design, cost, and capabilities of the SSBN(X)—the submarine class slated to replace the Ohio class—are among the most significant uncertainties in the Navy's and CBO's analyses of the cost of future shipbuilding....
>
> The recent history of cost estimates for the SSBN(X) illustrates both the high expected costs of the program and the uncertainty regarding those costs. The Navy's 2007 and 2008 shipbuilding plans included a projection that the SSBN(X) would cost an average of $3.8 billion (in 2013 dollars) per ship. The 2011 plan estimated the costs of the SSBN(X) class at an average of $7.9 billion apiece, while under the 2012 plan, the cost was lowered to $6.7 billion. The Navy currently estimates the cost of the lead SSBN(X) at $12.0 billion. The estimated average cost of follow-on ships is now $5.9 billion, and the Navy has stated an objective of reducing that cost to $5.4 billion in 2013 dollars. All told, the Navy estimates that building 12 submarines will cost $77 billion, an average of $6.4 billion each.
>
> Between the 2011 and 2012 plans, the Navy redefined its SSBN(X) design with the primary goal of reducing the cost. The Navy's cost estimate in the 2011 plan was based on a design similar in size to the Ohio class and on the cost of building Ohio class submarines using contemporary technology and under current conditions of the shipbuilding industry (such as the number of shipbuilders and vendors and the amount of other business in the shipyards). The Navy states that it was able to reduce the estimated cost of the SSBN(X) to the current projection by making the following changes:
>
> — Using a less expensive and more specific basic design (eliminating some costs in the estimate for the 2011 plan that were associated with uncertainty);

— Reducing the number of missile tubes from 20 to 16;

— Reducing the diameter of the missile tubes from 97 inches to 87 inches, which is the minimum needed to launch the Trident D-5 submarine-launched ballistic missile;

— Reducing the capability of the torpedo room and various sensor arrays and reducing the size of the sail mast;

— Increasing the use of components from the Virginia class attack submarines; and

— Simplifying many small elements in the design of the new submarine.

While the Navy estimates that the lead SSBN(X) will cost $12.0 billion, CBO estimates that it will cost $13.0 billion. Estimating the cost of the first submarine of a class is particularly difficult because it is not clear how much the Navy will spend on nonrecurring engineering and detailed design. The Navy spent about $2 billion on those items for the lead Virginia class attack submarine. The historical record for the lead ship of new classes of submarines in the 1970s and 1980s indicates that there is little difference in those items on a per-ton basis between a lead attack submarine and a lead SSBN. Therefore, CBO projects that the cost of nonrecurring engineering and detailed design is proportional to the weight of submarines, which implies that nonrecurring items would cost about $5 billion for the lead SSBN(X)—a submarine that will be somewhat larger than an Ohio class submarine and about 2½ times the size of a Virginia class submarine. The Navy's estimate for the lead SSBN(X) takes into account nonrecurring costs of an estimated $4.5 billion.

All told, 12 SSBN(X)s would cost about $87 billion in CBO's estimation, or an average of $7.2 billion each—$0.8 billion higher per boat than the Navy's estimate. That average includes the $13.0 billion estimated cost of the lead submarine and a $6.7 billion average estimated cost for the 2nd through 12th submarines. Research and development would cost an additional $10 billion to $15 billion, for a total program cost of $97 billion to $102 billion, CBO estimates.[68]

Program Affordability and Impact on Other Navy Shipbuilding Programs

Overview

Another oversight issue for Congress concerns the prospective affordability of the Ohio replacement program and its potential impact on funding available for other Navy shipbuilding programs. Even with the Navy's current effort to reduce the estimated unit procurement cost of the SSBN(X) toward DOD's target figure, observers are concerned that the Ohio replacement program could crowd out funding for other Navy shipbuilding programs in the 2020s and early 2030s. The Navy's report on the FY2014 30-year shipbuilding plan states:

> The Department [of the Navy] will encounter several challenges in executing this shipbuilding plan; perhaps the most important is funding and delivering the *Ohio*-replacement (OR) program SSBN. The OR SSBN is projected to cost about $6 billion each

---

[68] Congressional Budget Office, *An Analysis of the Navy's Fiscal Year 2014 Shipbuilding Plan*, October 2013, pp. 23-24.

[in constant FY2013 dollars]. Therefore, during the procurement and construction of OR SSBN between FY2021 and FY2035 an average of $19.2 billion per year is projected to be required for shipbuilding, which will be a key resourcing challenge for the Department....

The cost of the OR SSBN is significant relative to the [annual ship procurement] resources available to DoN in any given year. At the same time, the Department will have to address the block retirement of ships procured in large numbers during the 1980s which are reaching the end of their service lives. The confluence of these events prevents DoN from being able to shift resources within the shipbuilding account to accommodate the cost of the OR SSBN.

If DoN funds the OR SSBN from within its own resources, OR SSBN construction will take away from construction of other ships in the battle force such as attack submarines, destroyers, aircraft carriers and amphibious warfare shps. The resulting battle force will not meet the requirements of the FSA [the Force Structure Assessment that led to the Navy's current goal for achieving and maintianing a 306-ship fleet] and will therefore not be sufficient to implement the DSG [Defense Strategic Guidance]. In addition there will be significant impact to the shipbuilding industrial base.[69]

A May 2, 2013, press report states:

[Vice Admiral William Burke, Deputy Chief of Naval Operations for Warfare Systems], who is set to retire in the next few weeks,[70] spoke frankly about the undersea portion of the U.S. strategic nuclear triad "and its intersection with our shipbuilding plan."

His conclusion: "If we buy the [the 12 planned Ohio replacement (SSBNX) ballistic missile submarines] within existing [Navy] funds, we will not reach 300 ships. In fact, we'll find ourselves closer to 250. At these numbers, our global presence will be reduced such that we'll only be able to visit some areas of the world episodically."

Sequestration will only make the situation worse. Burke said it would cause the Navy "to both reduce procurement as well as retire existing ships, leaving us with a Navy in the vicinity of 200 ships, at which point we may not be considered a global navy."[71]

September 2013 Navy Testimony

As noted earlier, on September 18, 2013, Admiral Jonathan Greenert, the Chief of Naval Operations, testified that the Ohio replacement program "is the top priority program for the Navy." Greenert made the statement as part of a discussion of implications for Navy programs if DOD spending were reduced to the revised cap levels (i.e., the lower caps) in the Budget Control Act. In such a budget scenario, Greenert testified, "We would still be able to sustain today's ballistic missile submarine (SSBN) force. The SSBN(X) would still deliver in 2030 to replace retiring Ohio class SSBN while meeting requirements for SSBN presence and surge. This is the top priority program for the Navy."[72]

---

[69] *Report to Congress on the Annual Long-Range Plan for Construction of Naval Vessels for FY2014*, May 2013, pp. 11, 18-19.

[70] Vice Admiral Burke retired on May 20, 2013.

[71] Walter Pincus, "Budget Cuts Could Reshape Country's Ship Supply," *Washington Post*, May 2, 2013: 15.

[72] Statement of Admiral Jonathan Greenert, U.S. Navy, Chief of Naval Operations, Before the House Armed Services Committee on Planning for Sequestration in FY 2014 and Perspectives of the Military Services on the Strategic Choices and Management Review, September 18, 2013, p. 10.

Prior to September 2013, Navy officials had suggested that if the Navy does not receive additional funding to help pay for the Ohio replacement program, the Navy might need to reduce funding for other shipbuilding programs. At a September 12, 2013, hearing before the Seapower and Projection Forces subcommittee of the House Armed Services Committee on undersea warfare, a Navy official made this point more definitively, stating:

> The CNO has stated, his number one priority as the chief of Naval operations, is our—our strategic deterrent—our nuclear strategic deterrent. That will trump all other vitally important requirements within our Navy, but if there's only one thing that we do with our ship building account, we—we are committed to sustaining a two ocean national strategic deterrent that protects our homeland from nuclear attack, from other major war aggression and also access and extended deterrent for our allies.[73]

At this same hearing, Navy officials also testified more specifically than they have in the past on the amount of supplemental funding they are seeking for the Ohio replacement program, and on the potential consequences for other shipbuilding programs if this funding is not received. The Navy testified that the service is seeking about $4 billion per year over 15 years in supplemental funding—a total of about $60 billion—for the Ohio replacement program.[74] The 15 years in question, Navy officials suggested in their testimony, are the years in which the Ohio replacement boats are to be procured (FY2021-FY2035, as shown in **Table 2**).[75] The $60 billion in additional funding equates to an average of $5 billion for each of the 12 boats, which is close to the Navy's target of an average unit procurement cost of $4.9 billion in constant FY2010 dollars for boats 2-12 in the program. The Navy stated at the hearing that the $60 billion in supplemental funding that the Navy is seeking would equate to less than 1% of DOD's budget over the 15-year period. The Navy also suggested that the 41 pre-Ohio class SSBNs that were procured in the 1950s and 1960s (see **Table 1**) were partially financed with funding that was provided as a supplement to the Navy's budget.[76]

The Navy officials stated at the September 12 hearing that if the Navy were to receive about $30 billion in supplemental funding for the Ohio replacement program—about half the amount that the Navy is requesting—then the Navy would need to eliminate from its 30-year shipbuilding plan a notional total of 16 other ships, including, notionally, four Virginia-class attack submarines, four destroyers, and eight other combatant ships (which might mean ships such as

---

[73] Transcript of hearing. (Spoken remarks of Rear Admiral Richard Breckenridge. The other witness at the hearing was Rear Admiral David Johnson).

[74] Transcript of hearing. (Spoken remarks of Rear Admiral Richard Breckenridge.)

[75] Transcript of hearing. (Spoken remarks of Rear Admiral Richard Breckenridge.)

[76] Transcript of hearing (Spoken remarks of Rear Admiral Richard Breckenridge.) Regarding supplemental funding for the 41 earlier SSBNs, Breckenridge stated:

> The—just a little backstep and history to talk about the two other times that we've had to, as a nation, build the strategic deterrent. So in—in the '60s we built 41 SSBNs; they were called the 41 For Freedom. We did that in a seven-year period, which again is just an incredible—only in America could you go ahead and put out 41 ballistic missile submarines in a seven-year period.

> There was an impact to other shipbuilding accounts at that time, but the priority was such for national survival that we had to go ahead and—and make that a—an imperative and a priority.

> There was a supplement to the Navy's top line at that time when we—when we fielded the class, but it did leave—cast quite a shadow over the rest of the shipbuilding in the '60s.

> We recapitalized those 41 For Freedom with 18 Ohio-class SSBNs in the '80s. It was the Reagan years. There was a major naval buildup. And underneath the umbrella of that buildup we were able to afford as a nation the recapitalization of building 18 SSBNs.

Littoral Combat Ships or amphibious ships). Navy officials stated, in response to a question, that if the Navy were to receive none of the supplemental funding that it is requesting, then these figures could be doubled—that is, that the Navy would need to eliminate from its 30-year shipbuilding plan a notional total of 32 other ships, including, notionally, eight Virginia-class attack submarines, eight destroyers, and 16 other combatant ships.[77]

Some Options for Addressing the Issue

In addition to making further changes and refinements in the design of the SSBN(X), options for reducing the cost of the Ohio replacement program and the program's potential impact on funding available for other Navy shipbuilding programs include the following:

- using a joint block buy contract that would cover both the Ohio replacement program and the Virginia-class attack submarine program;

- altering the schedule for procuring the SSBN(X)s so as to create additional opportunities for using incremental funding for procuring the ships; and

- funding the procurement of SSBN(X)'s outside the Navy's shipbuilding budget; and

- reducing the planned number of SSBN(X)s.

Each of these options is discussed below.

Joint Block Buy Contract Covering Both Ohio Replacement and Virginia-Class Programs

To help reduce ship procurement costs, the Navy in recent years has made extensive use of multiyear procurement (MYP) contracts and block buy contracts in its shipbuilding programs,[78] including the Virginia class attack submarine program.[79] In light of this, the Navy will likely seek to use block buy and/or MYP contracting in the Ohio replacement program. More particularly, the Navy is investigating the possibility of using a single, joint-class block buy contract that would cover both Ohio replacement boats and Virginia class boats.[80] Such a contract, which could be viewed as precedent-setting in its scope, could offer savings beyond what would be possible using separate MYP or block buy contracts for the two submarine programs. A March 2014 GAO

---

[77] Transcript of hearing. (Spoken remarks of Rear Admiral Richard Breckenridge.) See also Christopher J. Castelli, "Admiral: DOD Likely To Support SSBN(X) Supplemental Funding," *Inside the Navy*, November 11, 2013.

[78] For more on MYP and block buy contracting, see CRS Report R41909, *Multiyear Procurement (MYP) and Block Buy Contracting in Defense Acquisition: Background and Issues for Congress*, by Ronald O'Rourke and Moshe Schwartz

[79] See CRS Report RL32418, *Navy Virginia (SSN-774) Class Attack Submarine Procurement: Background and Issues for Congress*, by Ronald O'Rourke.

[80] Government Accountability Office, *Defense Acquisitions[:] Assessments of Selected Weapons Programs*, GAO-14-340SP, March 2014, p. 141. See also Lee Hudson, "Navy leadership Continues To Rally Congress To Fund Sub Programs," *Inside the Navy*, February 17, 2014; Lee Hudson, "Navy Would Look To Cut VA-Class Sub Build Rate To Pay For SSBN(X)," Inside the Navy February 3, 2014; Lee Hudson, "Lower Ohio-Replacement Cost Tied To VA-Class Multiyear Deal," Inside the navy, May 20, 2013; Jason Sherman, "Navy Eyes Consolidation Of Sub Buys For Ohio Replacement, Virginia Class," *Inside the Navy*, February 28, 2011.

---

report states that if the Navy decides to propose such a contract, it would develop a legislative proposal in 2017.[81]

Altering Procurement Schedule to Make More Use of Incremental Funding

Another option for managing the potential impact of the Ohio replacement program on other Navy shipbuilding programs would be to stretch out the schedule for procuring SSBN(X)s and make greater use of split funding (i.e., two-year incremental funding) in procuring them.[82] This option would not reduce the total procurement cost of the Ohio replacement program—to the contrary, it might increase the program's total procurement cost somewhat by reducing production learning curve benefits in the Ohio replacement program.[83] This option could, however, reduce the impact of the Ohio replacement program on the amount of funding available for the procurement of other Navy ships in certain individual years. This might reduce the amount of disruption that the Ohio replacement program causes to other shipbuilding programs in those years, which in turn might avoid certain disruption-induced cost increases for those other programs. The annual funding requirements for the Ohio replacement program might be further spread out by funding some of the SSBN(X)s with three- or four-year incremental funding.[84]

Table 4 shows the Navy's currently planned schedule for procuring 12 SSBN(X)s and a notional alternative schedule that would start two years earlier and end two years later than the Navy's currently planned schedule. Although the initial ship in the alternative schedule would be procured in FY2019, it could be executed as it if were funded in FY2021. Subsequent ships in the alternative schedule that are funded earlier than they would be under the Navy's currently planned schedule could also be executed as if they were funded in the year called for under the Navy's schedule. Congress in the past has funded the procurement of ships whose construction was executed as if they had been procured in later fiscal years.[85] The ability to stretch the end of the procurement schedule by two years, to FY2035, could depend on the Navy's ability to carefully husband the use of the nuclear fuel cores on the last two Ohio-class SSBNs, so as to extend the service lives of these two ships by one or two years. Alternatively, Congress could grant the Navy the authority to begin construction on the 11th boat a year before its nominal year of procurement, and the 12th boat two years prior to its nominal year of procurement.

---

[81] Government Accountability Office, *Defense Acquisitions[:] Assessments of Selected Weapons Programs*, GAO-14-340SP, March 2014, p. 141.

[82] Under split funding, a boat's procurement cost is divided into two parts, or increments. The first increment would be provided in the fiscal year that the boat is procured, and the second would be provided the following fiscal year.

[83] Procuring one SSBN(X) every two years rather than at the Navy's planned rate of one per year could result in a loss of learning at the shipyard in moving from production of one SSBN to the next.

[84] The Navy, with congressional support, currently uses split funding to procure large-deck amphibious assault ships (i.e., LHAs). The Navy currently is permitted by Congress to use four-year incremental funding for procuring the first three Ford (CVN-78) class carriers (i.e., CVN-78, CVN-79, and CVN-80); the authority was granted in §121 of the FY2007 defense authorization act [H.R. 5122/P.L. 109-364 of October 17, 2006]).

[85] Congress funded the procurement of two aircraft carriers (CVNs 72 and 73) in FY1983, and another two (CVNs 74 and 75) in FY1988. Although CVN-73 was funded in FY1983, it was built on a schedule consistent with a carrier funded in FY1985; although CVN-75 was funded in FY1988, it was built on a schedule consistent with a carrier funded in FY1990 or FY1991.

Table 4. Navy SSBN(X) Procurement Schedule and a Notional Alternative Schedule

Fiscal year	Navy's Schedule	Boat might be particularly suitable for 2-, 3-, or 4-year incremental funding	Notional alternative schedule	Boat might be particularly suitable for 2-, 3-, or 4-year incremental funding
2019			I	X
2020				
2021	I	X	I	X
2022				
2023			I	X
2024	I	X		
2025			I	X
2026	I			
2027	I		I	
2028	I		I	
2029	I		I	
2030	I		I	
2031	I		I	X
2032	I			
2033	I	X	I	X
2034	I	X		
2035	I	X	I	X
2036				
2037			I	X
Total	**12**		**12**	

Source: Navy's current plan is taken from the Navy's FY2015 budget submission. Potential alternative plan prepared by CRS.

Notes: Notional alternative schedule could depend on Navy's ability to carefully husband the use of the nuclear fuel cores on the last two Ohio-class SSBNs, so as to extend the service lives of these two ships by one or two years. Alternatively, Congress could grant the Navy the authority to begin construction on the 11th boat a year before its nominal year of procurement, and the 12th boat two years prior to its nominal year of procurement. Under Navy's schedule, the boat to be procured in FY2033 might be particularly suitable for 4-year incremental funding, and boat to be procured in FY2034 might be particularly suitable for 3- or 4-year incremental funding.

A December 19, 2011, press report states:

> The Office of Management and Budget's Nov. 29[, 2011,] passback memorandum to the Defense Department [regarding the FY2013 DOD budget] warns that the effort to build replacements for aging Ohio-class submarines is not exempt from rules requiring each new vessel to be fully funded in a single year....
>
> Spreading the cost of a big-ticket ship over more than one year—an approach referred to as "incremental funding"—is only allowed when a program meets three criteria, OMB writes....
>
> "OMB does not anticipate that the OHIO Replacement program will meet these criteria," the passback memo states.[86]

---

[86] Christopher J. Castelli, "White House Opposes Incremental Funding For Multibillion-Dollar Sub," *Inside the Navy*, December 19, 2011.

Procuring SSBN(X)s Outside Navy's Shipbuilding Budget

Some observers have suggested funding the procurement of SSBN(X)s outside the Navy's shipbuilding budget, so as to preserve Navy shipbuilding funds for other Navy shipbuilding programs. There would be some precedent for such an arrangement:

- DOD sealift ships and Navy auxiliary ships are funded in the National Defense Sealift Fund (NDSF), a part of DOD's budget that is outside the Shipbuilding and Conversion, Navy (SCN) appropriation account, and also outside the procurement title of the DOD appropriations act.

- Most spending for ballistic missile defense (BMD) programs (including procurement-like activities) is funded through the Defense-Wide research and development account rather than through the research and development and procurement accounts of the individual military services.

A rationale for funding DOD sealift ships in the NDSF is that DOD sealift ships perform a transportation mission that primarily benefits services other than the Navy, and therefore should not be forced to compete for funding in a Navy budget account that funds the procurement of ships central to the Navy's own missions. A rationale for funding BMD programs together in the Defense-Wide research and development account is that this makes potential tradeoffs in spending among various BMD programs more visible and thereby helps to optimize the use of BMD funding.

As a reference tool for better understanding DOD spending, DOD includes in its annual budget submission a presentation of the DOD budget reorganized into 11 program areas, of which one is strategic forces. The FY2015 budget submission, for example, shows that about $11.7 billion is requested for strategic forces for FY2015.[87]

Supporters of funding the procurement of SSBN(X)s outside the Navy's shipbuilding budget might argue that this could help protect funding for other Navy shipbuilding programs. They could also argue that creating a new budget account for strategic nuclear forces of all kinds could help DOD better view potential tradeoffs in spending for various strategic nuclear forces programs and thereby help DOD better optimize the use of strategic forces funding.

Skeptics of funding the procurement of SSBN(X)s outside the Navy's shipbuilding budget could argue that it might do little to protect funding for other Navy shipbuilding programs, because if DOD were to move the SSBN(X)s out of the Navy's shipbuilding budget, DOD might also remove the funding that was there for the SSBN(X)s. They might also argue that shifting SSBN(X)s out of the Navy's shipbuilding budget would make it harder to track and maintain oversight over Navy shipbuilding activities, and that creating a new budget account for strategic nuclear forces of all kinds could endanger the Ohio replacement program by making it more visible to those who might support reduced spending on nuclear-weapon-related programs.

---

[87] Department of Defense, *National Defense Budget Estimates For FY 2015,* April 2014, Table 6-4, "Department of Defense TOA by Program," page 98. See also Table 6-5 on page 100, which presents the same data in constant FY2015 dollars. The other 10 program areas in addition to strategic forces are general purpose forces; C3, intelligence and space; mobility forces; guard and reserve forces; research and development; central supply and management; training, medical and other; administration and associated; support of other nations; and special operations forces. (A 12th category—other—shows relatively small amounts of funding.)

A March 11, 2010, press report stated: "The massive cost of replacing the Navy's nuclear ballistic missile submarines will be shouldered in the coming years by diverting funds from other naval and Pentagon programs and perhaps by boosting the defense budget, but the program should not get its own special funding stream, according to Deputy Defense Secretary William Lynn."[88]

A March 28, 2011, press report stated that SSBN(X)s

> will be funded within the shipbuilding account, not in a separate account as the sea service's top admiral has advocated, according to Pentagon acquisition chief Ashton Carter.
>
> "It's been in the shipbuilding account and our plan is it's going to stay in the shipbuilding account," Carter told *Inside the Pentagon* March 21 in a brief interview. "We just have to make it so that it is digestible for the Navy in the context of other shipbuilding needs. And we want the same things they want. We can manage through that path for decades."[89]

At an April 13, 2011, hearing on DOD acquisition before the Defense subcommittee of the House Appropriations Committee, the following exchange took place:

> REPRESNTATIVE CRENSHAW: Dr. Carter, I want to ask about Abrams tanks, kind of the modification of the start and stop. But – but real quick, we – we talked about the ballistic missiles submarines and was encouraged to hear that we've got a handle on the cost. We spent a lot of money on the development. I think we start construction in 2019.
>
> But even – even if we – you end up with a boat that costs $5 billion and we have 12 of those, that's $60 billion. And we talked about the difficult choices that's going to present in terms of surface ships, I just want to pose the question, if – is it under consideration to consider those submarines like a national asset?
>
> For instance, we – we fund the ballistic missile defense outside of the budget of the services because it's truly a national asset. And I wondered, it's a lot of money. And – and it's – those – those submarines are one-third of our nuclear triad. Is consideration being given to consider those being funded as a national asset outside the ship-building program which would take away some of the difficult choice in terms of the service ships versus the submarines?
>
> ASHTON CARTER, UNDER SCRETARY OF DEFENSE FOR ACQUISITON, TECHNOLOLGY, AND LOGISTICS: The – the best I can do is cite something that Secretary Gates said which is that he had considered that, then was not attracted to that idea. I'm paraphrasing, but I think their basic reason was, "Look, the money is going to show up somewhere anyway. And we're not going to hide $60 billion by re-labeling. So, let's keep it in a way we've – we've done it."
>
> And I think it was the gist of the secretary's response. So – so, Secretary Gates had considered it and was not attracted to the idea. Although he – he thoroughly recognizes the premise of your question which is there's a lot of money. And as a practical matter it will compete with those things in the defense budget.

---

[88] Christopher J. Castelli, "Lynn: Navy, DOD To Shoulder SSBN(X) Cost Without Separate Fund," *Inside the Pentagon*, March 11, 2010.

[89] Christopher J. Castelli, "Carter: Multibillion-Dollar Nuclear Subs Will Stay In Shipbuilding Account," *Inside the Navy*, March 28, 2011.

And that's one of the reasons why we've got to get the cost down.[90]

An August 1, 2011, press report stated:

> [Admiral Jonathan Greenert, who became the Chief of Naval Operations in September 2011, told] Sen. Jack Reed (D-R.I.) discussions are still underway in the Pentagon to have the defense-wide budget share with the Navy some of the costs of the Ohio-class SSBN(X) next-generation ballistic missile submarine, which is projected to dominate the Navy's shipbuilding budget starting at the end of this decade. "If confirmed, I intend to try to continue those discussions," Greenert [said] during his confirmation hearing. "In the [2020s], we have a phenomenon, an unfortunate one, where many of the ships built in the [1980s] will now come due for retirement. That's right when the Ohio replacement comes in. So we'll work very hard to make sure we got the requirements right. We'll work very hard with the acquisition community to drive that cost down but we may even so need some assistance, I believe, in the shipbuilding budget if we're going to meet our goals."[91]

At a March 29, 2012, hearing before the Senate Armed Services Committee to consider the nominations of several people for various DOD positions, the following exchange occurred:

> SENATOR JACK REED: Secretary Kendall, one of the issues that we have talked about is the nuclear infrastructure to create and maintain nuclear devices. There is another big part of that. That is the delivery platforms. And where you are facing a significant set of challenges, the lead procurement item is the Ohio class replacement submarine, but the Air Force is talking about the need ultimately to replace their fleet. You have to make, I presume, improvements in ground-based systems.
>
> When the services look individually at the cost—and I have got more fidelity with respect to the Navy—these are very, very expensive platforms. They crowd out spending for other necessary ships in the Navy's case. And I think there is a very compelling case because this is a strategic issue that the services alone should not fundamentally share the burden, that in fact there has to be some DOD defense money because of the strategic nature committed to help the services. And I think the most immediate situation is in the Navy.
>
> Can you reflect on that and share your views?
>
> Mr. FRANK KENDALL III.[92] Yes, Senator Reed. The Department [of Defense] basically builds its budget as a budget for the entire Department, and we do make tradeoffs that sometimes cut across the Services? [sic] lines in order to do that. Last fall, what we went through was a period where we formulated the strategy, the Strategic Guidance that we published, and that was used to guide the budget process. So that was all done with regard to priorities to support the strategy. It was not about the service portfolio specifically. At the end, we came to a decision about the best mix of systems to do that, and we tried to take into account the long-term issues that you alluded to which include the 30-year shipbuilding plan which we just sent over which does show that the Ohio replacement does add substantially to that account. We are going to have to find some other way besides the shipbuilding account obviously to pay that bill.

---

[90] Source: Transcript of hearing.

[91] "Boomer Sharing," *Defense Daily*, August 1, 2011: 1-2.

[92] Kendall at the time of the hearing was nominated for the position of Under Secretary of Defense for Acquisition, Technology and Logistics (USD ATL)—the DOD acquisition executive. He was subsequently confirmed for the position.

We have put cost caps on both the SSBN–X, the Ohio replacement, and on the new bomber in order to try to control the costs and keep them within an affordable range. But there is going to be a challenge to us to do this, and it has to be done on a defense-wide DOD basis.[93]

At an April 24, 2013, hearing before the Defense subcommittee of the Senate Appropriations Committee, the following exchange occurred:

SENATOR JACK REED (continuing): Let me just shift to another program which is critical to our national security. That's the Ohio-class replacement. And in fact, given its recognized ability to avoid detection, its invulnerability, it becomes more and more critical to the triad. And I'm wondering, Mr. Secretary and CNO, if you can comment on its growing importance in terms of the—of the need for it at sea, the replacement?

SECRETARY OF THE NAVY RAY MABUS: Well, that need has been amply documented, justified. We are on track with all the R&D and early development work to begin construction in 2021 for the first boat to put to sea in 2029, which would be exactly on schedule. We're also working very closely with—with our partners, the British, on the common missile compartment, since they—they are buying for their successor class the same missile compartment using the same missiles.

But the—one word of caution. We are on track today. It's a large program. It's an expensive program. And actually two words of caution. One is sequestration holds the potential to—to upset this timeline in a fairly dramatic way.

And second, as we get closer to time, we—there will have to be, as I believe Deputy Secretary of Defense Carter said in his transmittal of the shipbuilding report last year, a discussion in terms of the Ohio-class replacement and the rest of our shipbuilding programs in terms of how we finance this. Because for a period of time there, building these Ohio-class replacements, as I said, very expensive, incredibly important program, but we need to keep the rest of our shipbuilding base intact.

REED: If I can follow up, and Admiral Greenert, is there a possibility that if we slip this, it will have a point at which we cannot have as many ballistic missile submarines at sea as we need for deterrence, and for strategic posturing?

ADMIRAL JONATHAN GREENERT, CHIEF OF NAVAL OPERATIONS: That's feasible but unacceptable, I'd say, Senator. Yes, we—so we can't slip it. And the secretary had it right. People ask me what is my number-one program of concern, and I would tell you it's the Ohio replacement program. I look at that more than any other one.

REED: Well thank you, Admiral. Just one point is that this is part of a strategic—in fact, I would say the—the most survivable leg of the triad. And it—it's not just a Navy program. It's a national program. And I wonder if there's any consideration of supporting the Navy's efforts with—with funds that are more generically defense rather than more specifically Navy.

MABUS: I think that was the conversation I was referring to.[94]

At a May 8, 2013, hearing on Navy shipbuilding programs before the Seapower subcommittee of the Senate Armed Services Committee, the following exchange occurred:

---

[93] Transcript of hearing.

[94] Transcript of hearing.

SENATOR JACK REED, CHAIRMAN: Senator McCain pointed out, I think, appropriately that we've got several—many challenges. One is to have an affordable SSBN replacement for the Ohio class and the other is to maintain carrier production better at a level that we can afford.

With respect to the Ohio class specifically replacement, since it is a strategic asset because of its contact with the—it's part of a triad, are there any attempts to provide—supplement a funding to the Navy shipbuilding budget because of the strategic dimension or are those talks progressed or are they have been undertaken in DOD?

SEAN STACKLEY, ASSISTANT SECRETARY OF THE NAVY FOR RESEARCH, DEVELOPMENT, AND ACQUISITION: Sir, I can answer straightly as those talks have not progressed. (Inaudible) of that.

REED: OK, that's an interesting comment. Thank you very much.

STACKLEY: Let me go ahead and expand then.

REED: Yeah.

STACKLEY: The...

REED: That's a wrap. That was...

STACKLEY: Well, the Navy's plan in the fit-up [sic: FYDP], we think that the budget that we have assigned to the numbers that we plan on procuring in the fit-up [sic: FYDP] is within our reach if you park sequestration momentarily.

But when you get outside of the fit-up [FYDP], now you're quickly answering into the period where the Ohio Replacement dominates our shipbuilding plan. We spent a lot of effort over the last couple of years to go after the requirements to drive affordability through the requirements process and also through the design process. So it's something that started at about a $7 billion a unit cost for the Ohio Replacement.

Current estimate is in the $5.6 billion [range]. We are working through the design process to get it down with an objective of about $5 billion—$4.9 billion. That by itself does not bring the shipbuilding plan within the reach of affordability.

So if you look at that period of time and you look at the budget of forecast in that period of time, you have to go back to the period of the 80's when we are building up the 600-ship Navy to see those type of shipbuilding budget levels that are projected for the force that is laid out in the shipbuilding plan. And that is beyond our shipbuilding TOA by any method of extrapolation.

REED: And that is assuming that we can stabilize the cost in the other shipbuilding programs.

STACKLEY: Yes, Sir.[95]

At a September 12, 2013, hearing before the Seapower and Projection Forces subcommittee of the House Armed Services Committee on undersea warfare, a Navy official testified that

---

[95] Transcript of hearing.

... I do think it's important for the country to look at this [Ohio replacement effort] as a requirement above the Navy's [own requirements], [as] a strategic level requirement[,] and we ought to give it the gravity of attention and focus and insulation from the pressures of sequestration.

That said, the control of those resources must remain resident within the Navy with the control of our acquisition community. We know how to build submarines, we know how to oversee the building of submarines, [and] Electric Boat, (inaudible)[96] best submarine ship builders in the world.

We need to be able to make sure that if we come up with a creative, you know, strategic account [in the budget] for this [effort] that it's still the Navy and the ship building team that has the control and authority over those moneys as we—as we do this recapitalization to make it as affordable as possible.[97]

Reducing the Planned Number of SSBN(X)s

Some observers over the years have advocated or presented options for an SSBN force of fewer than 12 SSBNs. A November 2103 CBO report on options for reducing the federal budget deficit, for example, presented an option for reducing the SSBN force to eight boats as a cost-reduction measure.[98] Earlier CBO reports have presented options for reducing the SSBN force to 10 boats as a cost-reduction measure.[99] CBO reports that present such options also provide notional arguments for and against the options. A June 2010 report by a group known as the Sustainable Defense Task Force recommends reducing the SSBN force to 7 boats;[100] a September 2010 report from the Cato Institute recommends reducing the SSBN force to 6 boats,[101] and a September 2013 report from a group organized by the Stimson Center recommends reducing the force to 10 boats.[102]

Views on whether a force of fewer than 12 SSBN(X)s would be adequate could depend on, among other things, assessments of strategic nuclear threats to the United States and the role of SSBNs in deterring such threats as a part of overall U.S. strategic nuclear forces, as influenced by the terms of strategic nuclear arms control agreements.[103] Reducing the number of SSBNs below 12 could also raise a question as to whether the force should continue to be homeported at both Bangor, WA, and Kings Bay, GA, or consolidated at a single location.

---

[96] A press report indicates that this inaudible portion includes these words: "[and] Huntington Ingalls—the Navy's two submarine builders—are some of the." (Lee Hudson, "Navy Asks Congress To Set Up $60B Supplemental Fund For SSBN(X)," *Inside the Navy*, September 16, 2013.

[97] Transcript of hearing. (Spoken remarks of Rear Admiral Richard Breckenridge. The other witness at the hearing was Rear Admiral David Johnson.)

[98] Congressional Budget Office, *Options for Reducing the Deficit: 2014 to 2023*, November 2013, pp. 68-69.

[99] See, for example, Congressional Budget Office, *Rethinking the Trident Force*, July 1993, 78 pp.; and Congressional Budget Office, *Budget Options*, March 2000, p. 62.

[100] *Debt, Deficits, and Defense, A Way Forward[:] Report of the Sustainable Defense Task Force*, June 11, 2010, pp. 19-20.

[101] Benjamin H. Friedman and Christopher Preble, Budgetary Savings from Military Restraint, Washington, Cato Institute, September 23, 2010 (Policy Analysis No. 667), pp. 8.

[102] *Strategic Agility: Strong National Defense for Today's Global and Fiscal Realities*, Stimson, Washington, DC, 2013, p. 29. (Sponsored by the Peter G. Peterson Foundation, Prepared by Stimson, September 2013.)

[103] For further discussion, see CRS Report RL33640, *U.S. Strategic Nuclear Forces: Background, Developments, and Issues*, by Amy F. Woolf.

U.S. strategic nuclear deterrence plans require a certain number of strategic nuclear warheads to be available for use on a day-to-day basis. After taking into account warheads on the other two legs of the strategic nuclear triad, the number of warheads on an SSBN's SLBMs, and factors independent of the number of warheads on the SLBMs, this translates into a requirement for a certain number of SSBNs to be on station (i.e., within range of expected targets) in Pacific and Atlantic waters at any given moment. The SSBN force is sized to support this requirement. Given the time needed for at-sea training operations, restocking SSBNs with food and other consumables, performing maintenance and repair work on the SSBNs, and transiting to and from deterrent patrol areas, only a fraction of the SSBN force can be on patrol at any given moment. The Navy's position (see "Planned Procurement Quantity: 12 SSBN(X)s to Replace 14 Ohio-Class Boats" in "Background") is that the requirement for having a certain number of SSBNs on patrol at any given moment translates into a need for a force of 14 Ohio-class boats, and that this requirement can be met in the future by a force of 12 SSBN(X)s.

Construction Shipyard(s)

Another potential issue for Congress regarding the Ohio replacement program is which shipyard or shipyards would build SSBN(X)s. Two U.S. shipyards are capable of building nuclear-powered submarines—General Dynamics' Electric Boat Division (GD/EB) of Groton, CT, and Quonset Point, RI, and Newport News Shipbuilding (NNS), of Newport News, VA, which forms part of Huntington Ingalls Industries (HII). GD/EB's primary business is building nuclear-powered submarines; it can also perform submarine overhaul work. NNS's primary lines of business are building nuclear-powered aircraft carriers, building nuclear-powered submarines, and performing overhaul work on nuclear-powered aircraft carriers.

Table 5 shows the numbers of SSBNs built over time by GD/EB, NNS, and two government-operated naval shipyards (NSYs)—Mare Island NSY, located in the San Francisco Bay area, and Portsmouth NSY of Portsmouth, NH, and Kittery, ME. Mare Island NSY is no longer in operation. NSYs have not built new Navy ships since the early 1970s; since that time, they have focused solely on overhauling and repairing Navy ships.

Table 5. Construction Shipyards of U.S. SSBNs

	George Washington (SSBN-598) class	Ethan Allen (SSBN-608) class	Lafayette/ Benjamin Franklin (SSBN-616/640) class	Ohio (SSBN-726) class
Fiscal years procured	FY58-FY59	FY59 and FY61	FY61-FY64	FY77-FY91
Number built by GD/EB	2	2	13	18
Number built by NNS	1	3	10	
Number built by Mare Island NSY	1		6	
Number built by Portsmouth NSY	1		2	
Total number in class	**5**	**5**	**31**	**18**

Source: Prepared by CRS based on data in Norman Polmar, *The Ships and Aircraft of the U.S. Fleet*, Annapolis, Naval Institute Press, various editions. NSY means naval shipyard.

Notes: GD/EB was the builder of the first boat in all four SSBN classes. The George Washington-class boats were procured as modifications of SSNs that were already under construction. A total of 18 Ohio-class SSBNs were built; the first four were converted into SSGNs in 2002-2008, leaving 14 in service as SSBNs.

As can be seen in the table, the Ohio-class boats were all built by GD/EB, and the three previous SSBN classes were built partly by GD/EB, and partly by NNS. GD/EB was the builder of the first boat in all four SSBN classes. The most recent SSBNs built by NNS were the *George C. Marshall* (SSBN-654) and *George Washington Carver* (SSBN-656), which were Lafayette/Benjamin Franklin-class boats that were procured in FY1964 and entered service in 1966.

There are at least five basic possibilities for building SSBN(X)s:

- **build all SSBN(X)s at GD/EB**—the approach that was used for building the Ohio-class SSBNs;

- **build all SSBN(X)s at NNS**;

- **build some SSBN(X)s GD/EB and some at NNS**—the approach that was used for building the George Washington-, Ethan Allen-, and Lafayette/Benjamin Franklin-class SSBNs;

- **build each SSBN(X) jointly at GD/EB and NNS, with final assembly of the boats alternating between the yards**—the approach currently being used for building Virginia-class SSNs;[104] and

- **build each SSBN(X) jointly at GD/EB and NNS, with one yard—either GD/EB or NNS—performing final assembly on every boat**.

In assessing these five approaches, policy makers may consider a number of factors, including their potential costs, their potential impacts on employment levels at GD/EB and NNS, and the relative value of preserving SSBN-unique construction skills (such as those relating to the construction and installation of SLBM compartments) at one shipyard or two. The relative costs of these five approaches could depend on a number of factors, including the following:

- each yard's share of SSBN(X) production work (if both yards are involved);

---

[104] Under the joint-production arrangement for Virginia-class boats, GD/EB builds certain parts of each boat, NNS builds certain other parts of each boat, and the two yards take turns building the reactor compartment and performing final assembly work. GD/EB is the final assembly yard for the first Virginia-class boat, the third one, and so on, while NNS is the final assembly yard for the second boat, the fourth one, and so on. The arrangement provides a roughly 50-50 split in profits between the two firms for the production of Virginia-class SSNs. The agreement governing the joint-production arrangement cannot be changed without the consent of both firms. Virginia-class SSNs are the first U.S. nuclear-powered submarines to be built jointly by two shipyards; all previous U.S. nuclear-powered submarines were built under the more traditional approach of building an entire boat within a single yard.

The Virginia-class joint-production arrangement was proposed by the two shipyards, approved by the Navy, and then approved by Congress as part of its action on the FY1998 defense budget. A principal goal of the arrangement is to preserve submarine-construction skills at two U.S. shipyards while minimizing the cost of using two yards to build a class of submarines that is procured at a relatively low rate of one or two boats per year. Preserving submarine-construction skills at two yards is viewed as a hedge against the possibility of operations at one of the yards being disrupted by a natural or man-made disaster.

The joint-production arrangement is more expensive than single-yard strategy of building all Virginia-class boats at one shipyard (in part because the joint-production strategy splits the learning curve for reactor compartment construction and final assembly work on Virginia-class SSNs), but it is less expensive than a separate-yard strategy of building complete Virginia-class separately at both yards (in part because a separate-construction strategy splits the learning curve for all aspects of Virginia-class construction work, and because, in the absence of other submarine-construction work, a procurement rate of one or two Virginia-class boats per year is viewed as insufficient to sustain a meaningful competition between the two yards for contracts to build the boats).

- the number of SSNs procured during the years of SSBN(X) procurement (which can affect economies of scale in submarine production);

- whether the current joint-production arrangement for the Virginia class remains in effect during those years;[105] and

- the volume of non-submarine-construction work performed at the two shipyards during these years, which would include in particular aircraft carrier construction and overhaul work at NNS.

A January 12, 2011, press report stated:

> While the [SSBN(X)] submarine-building contracts would likely be competitively bid, [Electric Boat President John] Casey says he doubts any other company—even its attack-submarine-building partner Northrop Grumman [now NNS]—can secure the work. Electric Boat built the existing Ohio-class fleet.
>
> "We have every intention of building every one of those ships," he says. "There's no one else [who was] involved in designing and building that [Ohio-class] platform.[106] It's up to us to convince people we can do it at the right price."[107]

Legislative Activity for FY2015

Bills introduced in the 113th Congress that would, among other things, limit the Ohio replacement program to no more than eight boats include H.R. 505, H.R. 1506, H.R. 4107, and S. 2070.

FY2015 Funding Request

As shown in **Table 3**, the Navy's proposed FY2015 budget requests $1,219.3 million in research and development funding for the Ohio replacement program, including $370.0 million for Project 3219 (SSBN[X] reactor plant) in PE0603570N (Advanced Nuclear Power Systems), $812.8 million for Project 3220 (SBSD [Sea-Based Strategic Deterrent] Advanced Submarine System Development) in PE0603595N (Ohio Replacement), and $36.5 million for Project 3237 (Launch Test Facility) in PE0603595N (Ohio Replacement).

---

[105] The agreement governing the joint-production arrangement for the Virginia class cannot be changed without the consent of both yards.

[106] The bracketed words in this sentence were inserted by CRS following a February 8, 2011, telephone call to CRS from Electric Boat in which Electric Boat stated that sentence in Mr. Casey's quote refers to Electric Boat being the sole designer and builder of the current Ohio-class SSBNs.

[107] Michael Fabey, "Electric Boat Recruits Engineers For Ohio-Class Sub Replacement," *Aerospace Daily & Defense Report*, January 12, 2011: 1-2.

Appendix A. June 2013 Navy Blog Post Regarding Ohio Replacement Options

This appendix presents the text of a June 26, 2013, blog post by Rear Admiral Richard Breckenridge, the Navy's Director for Undersea Warfare (N97), discussing options that were examined for replacing the Ohio-class SSBNs. The text is as follows:

> Over the last five years, the Navy – working with U.S. Strategic Command, the Joint Staff and the Office of the Secretary of Defense – has formally examined various options to replace the Ohio ballistic missile submarines as they retire beginning in 2027. This analysis included a variety of replacement platform options, including designs based on the highly successful Virginia-class attack submarine program and the current Ohio-class ballistic missile submarine. In the end, the Navy elected to pursue a new design that leverages the lessons from the Ohio, the Virginia advances in shipbuilding and improvements in cost-efficiency.
>
> Recently, a variety of writers have speculated that the required survivable deterrence could be achieved more cost effectively with the Virginia-based option or by restarting the Ohio-class SSBN production line. Both of these ideas make sense at face value – which is why they were included among the alternatives assessed – but the devil is in the details. When we examined the particulars, each of these options came up short in both military effectiveness and cost efficiency.
>
> **Virginia-based SSBN design with a Trident II D5 missile.** An SSBN design based on a Virginia-class attack submarine with a large-diameter missile compartment was rejected due to a wide range of shortfalls. It would:
>
> • Not meet survivability (stealth) requirements due to poor hull streamlining and lack of a drive train able to quietly propel a much larger ship
>
> • Not meet at-sea availability requirements due to longer refit times (since equipment is packed more tightly within the hull, it requires more time to replace, repair and retest)
>
> • Not meet availability requirements due to a longer mid-life overhaul (refueling needed)
>
> • Require a larger number of submarines to meet the same operational requirement
>
> • Reduce the deterrent value needed to protect the country (fewer missiles, warheads at-sea)
>
> • Be more expensive than other alternatives due to extensive redesign of Virginia systems to work with the large missile compartment (for example, a taller sail, larger control surfaces and more robust support systems)
>
> We would be spending more money (on more ships) to deliver less deterrence (reduced at-sea warhead presence) with less survivability (platforms that are less stealthy).
>
> **Virginia-based SSBN design with a smaller missile.** Some have encouraged the development of a new, smaller missile to go with a Virginia-based SSBN. This would carry forward many of the shortfalls of a Virginia-based SSBN we just discussed, and add to it a long list of new issues. Developing a new nuclear missile from scratch with an industrial base that last produced a new design more than 20 years ago would be challenging, costly

and require extensive testing. We deliberately decided to extend the life of the current missile to decouple and de-risk the complex (and costly) missile development program from the new replacement submarine program. Additionally, a smaller missile means a shorter employment range requiring longer SSBN patrol transits. This would compromise survivability, require more submarines at sea and ultimately weaken our deterrence effectiveness. With significant cost, technical and schedule risks, there is little about this option that is attractive.

Ohio-based SSBN design. Some have argued that we should re-open the Ohio production line and resume building the Ohio design SSBNs. This simply cannot be done because there is no Ohio production line. It has long since been re-tooled and modernized to build state-of-the-art Virginia-class SSNs using computerized designs and modular, automated construction techniques. Is it desirable to redesign the Ohio so that a ship with its legacy performance could be built using the new production facilities? No, since an Ohio-based SSBN would:

•   Not provide the required quieting due to Ohio design constraints and use of a propeller instead of a propulsor (which is the standard for virtually all new submarines)

•   Require 14 instead of 12 SSBNs by reverting to Ohio class operational availability standards (incidentally creating other issues with the New START treaty limits)

•   Suffer from reduced reliability and costs associated with the obsolescence of legacy Ohio system components

Once again, the end result would necessitate procuring more submarines (14) to provide the required at-sea presence and each of them would be less stealthy and less survivable against foreseeable 21st century threats.

The Right Answer: A new design SSBN that improves on Ohio: What has emerged from the Navy's exhaustive analysis is an Ohio replacement submarine that starts with the foundation of the proven performance of the Ohio SSBN, its Trident II D5 strategic weapons system and its operating cycle. To this it adds:

•   Enhanced stealth as necessary to pace emerging threats expected over its service life

•   Systems commonality with Virginia (pumps, valves, sonars, etc.) wherever possible, enabling cost savings in design, procurement, maintenance and logistics

•   Modular construction and use of COTS equipment consistent with those used in today's submarines to reduce the cost of fabrication, maintenance and modernization. Total ownership cost reduction (for example, investing in a life-of-the-ship reactor core enables providing the same at-sea presence with fewer platforms). Although the Ohio replacement is a "new design," it is in effect an SSBN that takes the best lessons from 50 years of undersea deterrence, from the Ohio, from the Virginia, from advances in shipbuilding efficiency and maintenance, and from the stern realities of needing to provide survivable nuclear deterrence. The result is a low-risk, cost-effective platform capable of smoothly transitioning from the Ohio and delivering effective 21st century undersea strategic deterrence.[108]

---

[108] "Facts We Can Agree Upon About Design of Ohio Replacement SSBN," Navy Live, accessed July 3, 2013, at http://navylive.dodlive mil/2013/06/26/facts-we-can-agree-upon-about-design-of-ohio-replacement-ssbn/.

Appendix B. Earlier Oversight Issue:
A Design with 16 vs. 20 SLBM Tubes

Overview

An earlier oversight issue for Congress concerned the plan to design the SSBN(X) with 16 SLBM tubes rather than 20—one of several decisions made to reduce the estimated average procurement cost of boats 2 through 12 in the program to $5.6 billion in FY2010 dollars.[109] Some observers were concerned that designing the SSBN(X) with 16 tubes rather than 20 would create a risk that U.S. strategic nuclear forces might not have enough capability in the 2030s and beyond to fully perform their deterrent role. These observers noted that to comply with the New Start Treaty limiting strategic nuclear weapons, DOD plans to operate in coming years a force of 14 Trident SSBNs, each with 20 operable SLBM tubes (4 of the 24 tubes on each boat are to be rendered inoperable), for a total of 240 tubes, whereas the Navy in the Ohio replacement program is planning a force of 12 SSBNs each with 16 tubes, for a total of 192 tubes, or 20% less than 240. These observers also cited the uncertainties associated with projecting needs for strategic deterrent forces out to the year 2080, when the final SSBN(X) is scheduled to leave service. These observers asked whether the plan to design the SSBN(X) with 16 tubes rather than 20 was fully supported within all parts of DOD, including U.S. Strategic Command (STRATCOM).

In response, Navy and other DOD officials stated that the decision to design the SSBN(X) with 16 tubes rather than 20 was carefully considered within DOD, and that they believe a boat with 16 tubes will give U.S. strategic nuclear forces enough capability to fully perform their deterrent role in the 2030s and beyond.

---

[109] At a March 30, 2011, hearing before the Strategic Forces subcommittee of the Senate Armed Services Committee, Admiral Kirkland Donald, Deputy Administrator for Naval Reactors and Director, Naval Nuclear Propulsion, National Nuclear Security Administration, when asked for examples cost efficiencies that are being pursued in his programs, stated:

> The—the Ohio replacement [program] has been one that we've obviously been focused on here for—for several years now. But in the name of the efficiencies, and one of the issues as we work through the Defense Department's acquisition process, we were the first program through that new process that Dr. [Aston] Carter [the DOD acquisition executive] headed up.

> But we were challenged to—to drive the cost of that ship down, and as far as our part was concerned, one of the key decisions that was made that—that helped us in that regard was a decision to go from 20 missile tubes to 16 missile tubes, because what that allowed us to do was to down rate the—the propulsion power that was needed, so obviously, it's a – it's a small[er] the reactor that you would need.

> But what it also allowed us to do was to go back [to the use of existing components]. The size [of the ship] fell into the envelope where we could go back and use components that we had already designed for the Virginia class [attack submarines] and bring those into this design, not have to do it over again, but several of the mechanical components, to use those over again.

> And it enabled us to drive the cost of that propulsion plant down and rely on proven technology that's—pumps and valves and things like that don't change like electronics do.

> So we're pretty comfortable putting that in ship that'll be around 'til 2080. But we were allowed to do that.

> (Source: Transcript of hearing.)

Testimony in 2011

At a March 1, 2011, hearing before the House Armed Services Committee, Admiral Gary Roughead, then-Chief of Naval Operations, stated:

> I'm very comfortable with where we're going with SSBN-X. The decision and the recommendation that I made with regard to the number of tubes—launch tubes are consistent with the new START treaty. They're consistent with the missions that I see that ship having to perform. And even though it may be characterized as a cost cutting measure, I believe it sizes the ship for the missions it will perform.[110]

At a March 2, 2011, hearing before the Strategic Forces subcommittee of the House Armed Services Committee, the following exchange occurred:

REPRESENTATIVE TURNER:

General Kehler, thank you so much for your continued thoughts and of course your leadership. One item that we had a discussion on was the triad, of looking to—of the Navy and the tube reductions of 20 to 16, as contained in other hearings on the Hill today. I would like your thoughts on the reduction of the tubes and what you see driving that, how you see it affecting our strategic posture and any other thoughts you have on that?

AIR FORCE GENERAL C. ROBERT KEHLER, COMMANDER, U.S. STRATEGIC COMMAND

Thank you, Mr. Chairman. Well, first of all, sir, let me say that the—in my mind anyway, the discussion of Trident and Ohio-class replacement is really a discussion in the context of the need to modernize the entire triad. And so, first of all, I think that it's important for us to recognize that that is one piece, an important piece, but a piece of the decision process that we need to go through.

Second, the issue of the number of tubes is not a simple black-and-white answer. So let me just comment here for a minute.

First of all, the issue in my mind is the overall number of tubes we wind up with at the end, not so much as the number of tubes per submarine.

Second, the issue is, of course, we have flexibility and options with how many warheads per missile per tube, so that's another consideration that enters into this mixture.

Another consideration that is important to me is the overall number of boats and the operational flexibility that we have with the overall number of boats, given that some number will need to be in maintenance, some number will need to be in training, et cetera.

And so those and many other factors—to include a little bit of foresight here, in looking ahead to 20 years from now in antisubmarine warfare environment that the Navy will have to operate in, all of those bear on the ultimate sideways shape configuration of a follow-on to the Ohio.

---

[110] Source: Transcript of hearing.

At this point, Mr. Chairman, I am not overly troubled by going to 16 tubes. As I look at this, given that we have that kind of flexibility that I just laid out; given that this is an element of the triad and given that we have some decision space here as we go forward to decide on the ultimate number of submarines, nothing troubles me operationally here to the extent that I would oppose a submarine with 16 tubes.

I understand the reasons for wanting to have 20. I understand the arguments that were made ahead of me. But as I sit here today, given the totality of the discussion, I am—as I said, I am not overly troubled by 16. Now, I don't know that the gavel has been pounded on the other side of the river yet with a final decision, but at this point, I am not overly troubled by 16.[111]

At an April 5, 2011, hearing before the Strategic Forces subcommittee of the House Armed Services Committee, the following exchange occurred:

REPRESENTATIVE LARSEN:

General Benedict, we have had this discussion, not you and I, I am sorry. But the subcommittee has had a discussion in the past with regards to the Ohio-class replacement program.

The new START, though, when it was negotiated, assumed a reduction from 24 missile tubes per hole to, I think, a maximum a maximum of 20.

The current configuration [for the SSBN(X)], as I understand it, would move from 24 to 16.

Can you discuss, for the subcommittee here, the Navy's rationale for that? For moving from 24 to 16 as opposed to the max of 20?

NAVY REAR ADMIRAL TERRY BENEDICT, DIRECTOR, STRATEGIC SYSTEMS PROGRAMS (SSP):

Sir, as part—excuse me, as part of the work-up for the milestone A [review for the SSBN(X) program] with Dr. Carter in OSD, SSP supported the extensive analysis at both the OSD level as well as STRATCOM's analysis.

Throughout that process, we provided, from the SWS [strategic weapon system] capability, our perspective. Ultimately that was rolled up into both STRATCOM and OSD and senior Navy leadership and in previous testimony, the secretary of the Navy, the CNO, and General Chilton have all expressed their confidence that the mission of the future, given their perspectives, is they see the environment today can be met with 16.

And so, as the acquisition and the SWS provider, we are prepared to support that decision by leadership, sir.

REPRESENTATIVE LARSEN:

Yes.

And your analysis supports—did your analysis that fed into this, did you look at specific numbers then?

---

[111] Source: Transcript of hearing.

REARD ADMIRAL BENEDICT:

Sir, we looked at the ability of the system, again, SSP does not look at specific targets with...

REPRESENTATIVE LARSEN:

Right. Yes, yes, yes.

REAR ADMIRAL BENEDICT:

Our input was the capability of the missile, the number of re-entry bodies and the throw weight that we can provide against those targets and based on that analysis, the leadership decision was 16, sir.[112]

At an April 6, 2011, hearing before the Strategic Forces subcommittee of the Senate Armed Services Committee, the following exchange occurred:

SENATOR SESSIONS:

Admiral Benedict, according to recent press reports, the Navy rejected the recommendations of Strategic Command to design the next generation of ballistic missile submarines with 20 missile tubes instead of opting for only 16 per boat.

What is the basis for the Navy's decision of 16? And I'm sure cost is a factor. In what ways will that decision impact the overall nuclear force structure associated with the command?

NAVY REAR ADMIRAL TERRY BENEDICT, DIRECTOR, STRATEGIC SYSTEMS PROGRAMS (SSP):

Yes, sir. SSP supported the Navy analysis, STRATCOM's analysis, as well as the OSD analysis, as we proceeded forward and towards the Milestone A decision [on the SSBN(X) program] that Dr. Carter conducted.

Based on our input, which was the technical input as the—as the director of SSP, other factors were considered, as you stated. Cost was one of them. But as the secretary, as the CNO, and I think as General Kehler submitted in their testimony, that given the threats that we see today, given the mission that we see today, given the upload capability of the D-5, and given the environment as they saw today, all three of those leaders were comfortable with the decision to proceed forward with 16 tubes, sir.

SENATOR SESSIONS:

And is that represent your judgment? To what extent were you involved—were you involved in that?

REAR ADMIRAL BENEDICT:

Sir, we were involved from technical aspects in terms of the capability of the missile itself, what we can throw, our range, our capability. And based on what we understand the

---

[112] Source: Transcript of hearing.

capability of the D-5 today, which will be the baseline missile for the Ohio Replacement Program, as the director of SSP I'm comfortable with that decision.[113]

Section 242 Report

Section 242 of the FY2012 National Defense Authorization Act (H.R. 1540/P.L. 112-81 of December 31, 2011) required DOD to submit a report on the Ohio replacement program that includes, among other things, an assessment of various combinations of boat quantities and numbers of SLBM launch tubes per boat. The text of the section is as follows:

> SEC. 242. REPORT AND COST ASSESSMENT OF OPTIONS FOR OHIO-CLASS REPLACEMENT BALLISTIC MISSILE SUBMARINE.
>
> (a) Report Required- Not later than 180 days after the date of the enactment of this Act, the Secretary of the Navy and the Commander of the United States Strategic Command shall jointly submit to the congressional defense committees a report on each of the options described in subsection (b) to replace the Ohio-class ballistic submarine program. The report shall include the following:
>
> (1) An assessment of the procurement cost and total life-cycle costs associated with each option.
>
> (2) An assessment of the ability for each option to meet—
>
> (A) the at-sea requirements of the Commander that are in place as of the date of the enactment of this Act; and
>
> (B) any expected changes in such requirements.
>
> (3) An assessment of the ability for each option to meet—
>
> (A) the nuclear employment and planning guidance in place as of the date of the enactment of this Act; and
>
> (B) any expected changes in such guidance.
>
> (4) A description of the postulated threat and strategic environment used to inform the selection of a final option and how each option provides flexibility for responding to changes in the threat and strategic environment.
>
> (b) Options Considered- The options described in this subsection to replace the Ohio-class ballistic submarine program are as follows:
>
> (1) A fleet of 12 submarines with 16 missile tubes each.
>
> (2) A fleet of 10 submarines with 20 missile tubes each.
>
> (3) A fleet of 10 submarines with 16 missile tubes each.
>
> (4) A fleet of eight submarines with 20 missile tubes each.

---

[113] Source: Transcript of hearing.

(5) Any other options the Secretary and the Commander consider appropriate.

(c) Form- The report required under subsection (a) shall be submitted in unclassified form, but may include a classified annex.

Subsection (c) above states the report "shall be submitted in unclassified form, but may include a classified annex."

The report as submitted was primarily the classified annex, with a one-page unclassified summary, the text of which is as follows (underlining as in the original):

> The National Defense Authorization Act (NDAA) for Fiscal Year 2012 (FY12) directed the Secretary of the Navy and the Commander of U.S. Strategic Command (USSTRATCOM) to jointly submit a report to the congressional defense committees comparing four different options for the OHIO Replacement (OR) fleet ballistic missile submarine (SSBN) program. Our assessment considered the current operational requirements and guidance. The four SSBN options analyzed were:
>
> 1. 12 SSBNs with 16 missile tubes each
>
> 2. 10 SSBNs with 20 missile tubes each
>
> 3. 10 SSBNs with 16 missile tubes each
>
> 4. 8 SSBNs with 20 missile tubes each
>
> The SSBN force continues to be an integral part of our nuclear Triad and contributes to deterrence through an assured second strike capability that is survivable, reliable, and credible. The number of SSBNs and their combined missile tube capacity are important factors in our flexibility to respond to changes in the threat and uncertainty in the strategic environment.
>
> We assessed each option against the ability to meet nuclear employment and planning guidance, ability to satisfy at-sea requirements, flexibility to respond to future changes in the postulated threat and strategic environment, and cost. In general, options with more SSBNs can be adjusted downward in response to a diminished threat; however, options with less SSBNs are more difficult to adjust upward in response to a growing threat.
>
> Clearly, a smaller SSBN force would be less expensive than a larger force, but for the reduced force options we assessed, they fail to meet current at-sea and nuclear employment requirements, increase risk in force survivability, and limit flexibility in response to an uncertain strategic future. <u>Our assessment is the program of record, 12 SSBNs with 16 missile tubes each, provides the best balance of performance, flexibility, and cost meeting commander's requirements while supporting the Nation's strategic deterrence mission goals and objectives.</u>
>
> The classified annex contains detailed analysis that is not releasable to the public.[114]

---

[114] Report and Cost Assessment of Options for OHIO-Class Replacement Ballistic Missile Submarine, Unclassified Summary, received from Navy Legislative Affairs Office, August 24, 2012. See also Christopher J. Castelli, "Classified Navy Assessment On SSBN(X) Endorses Program Of Record," *Inside the Navy*, September 10, 2012.

Author Contact Information

Ronald O'Rourke
Specialist in Naval Affairs
rorourke@crs.loc.gov, 7-7610